T0205413

Cognitive Systems Monographs

Volume 38

The Cognitive Systems Monographs (COSMOS) publish new developments and advances in the fields of cognitive systems research, rapidly and informally but with a high quality. The intent is to bridge cognitive brain science and biology with engineering disciplines. It covers all the technical contents, applications, and multidisciplinary aspects of cognitive systems, such as Bionics, System Analysis, System Modelling, System Design, Human Motion, Understanding, Human Activity Understanding, Man-Machine Interaction, Smart and Cognitive Environments, Human and Computer Vision, Neuroinformatics, Humanoids, Biologically motivated systems and artefacts Autonomous Systems, Linguistics, Sports Engineering, Computational Intelligence, Biosignal Processing, or Cognitive Materials as well as the methodologies behind them. Within the scope of the series are monographs, lecture notes, selected contributions from specialized conferences and workshops.

More information about this series at http://www.springer.com/series/8354

David Israel González Aguirre

Visual Perception for Humanoid Robots

Environmental Recognition and Localization, from Sensor Signals to Reliable 6D Poses

David Israel González Aguirre
Software and Systems Research (SSR),
Anticipatory Computing Lab (ACL)
Intel Labs
Hillsboro, OR, USA

ISSN 1867-4925 ISSN 1867-4933 (electronic)
Cognitive Systems Monographs
ISBN 978-3-030-07415-9 ISBN 978-3-319-97841-3 (eBook)
https://doi.org/10.1007/978-3-319-97841-3

This Springer imprint is published by the registered company Springer Nature Switzerland AG
The registered company address is: Gewerbestrasse 11, 6330 Cham, Switzerland

To my parents

Acknowledgements

I express my sincerest gratitude to my supervisor Prof. Rüdiger Dillmann for all the support during these years and for giving me the opportunity to experience the wide spectrum of academic life at the KIT. I thank Prof. Jan-Olof Eklundh for kindly co-supervising my research work. All my gratitude to Prof. Tamim Asfour for all his patience, fostering, technical and scientific advice, without his enthusiasm and belief, this work would not exist.

Thanks to my colleagues for their support and advice in all aspects of the long work and fun. In particular, all my gratitude to Stefan Ulbrich for his time, effort and constructive criticism along the years. Thanks a lot to Paul Holz, Sebastian Schulz, Martin Do, Nikolaus Vahrenkamp, Julian Schill, Steven Wieland, Kai Welke, Alexander Bierbaum, Pedram Azad, Markus Przybylski, Christian Böge, Manfred Kröhnert, David Schiebener, Mirko Wächter, Ömer Terlemez, Pascal Meissner, Tobias Gindele, Sebastian Brechtel, Sebastian Röhl and Michael Neaga for all the motivation, help, support and productive collaboration in demonstrations, publications, seminars, experiments and lectures. Thanks to the Humanoids and Intelligence Systems Lab for their support and assistance. Additional thanks to Isabelle Wappler, Diana Kreidler, Christine Brand and Sandra Tartarelli.

I especially appreciated and enjoyed the opportunity to supervise the master and bachelor theses of Julian Hoch, Raoul Gabriel, Jonatan Dominguez and Michael Vollert. Thanks for your hard work and intensive discussions which allowed not only the successful achievement of your respective degrees but also creating important software components and visible publications. My sincere thanks to the scientific and technical enrichment resulting from the discussions with Prof. Eduardo Bayro-Corrochano, Julio Zamora-Esquivel, Miguel Bernal-Marin, José Luis Cuéllar-Camacho, Javier Felip-León, Lothar Felten and Sinisa Kontin.

I deeply thank Nathalie Grün for the unconditional belief along the hardest moments of this Ph.D., *Alle hier kochen auch nur mit Wasser*. Thanks for your words, support and friendship Domieque Grün, Guntram Grün, Linda Schabach, Thorsten Nickel, Sinisa Kontin, Ruben Aguirre, Jorge Romero and Anna Ninic.

Thanks a lot to my parents for providing me with all the emotional, spiritual and material foundation for this ongoing journey. Thanks to my brothers for sharing with me their knowledge, experience and courage to face and enjoy the challenges. Thanks to my uncle Guillermo for being a continuous source of support and motivation. In general to all my relatives whose support has given me a strong companion along my life.

Finally, I gratefully acknowledge the founding resources provided by the Ph.D. scholarship CONACYT-DAAD Nr. 180868. I express my gratitude for all the learning and founding during my participation in the German project SFB-588 and the European projects EU FP7 GRASP and EU FP7 ICT Xperience.

Contents

List of Figures

List of Tables

Chapter 1
Introduction

A journey of a thousand miles begins with a single step
Lao Tzu

Artificial systems—from the simple lever to the state-of-the-art super computers—have shaped and are strongly transforming the physical and social structure of the human kind. Emerging from this evolving scientific, technological and social process, there is a particular interesting field involving not only fascinating cutting-edge machines and sophisticated methods but also embracing the nature of the humankind itself, namely the research on anthropomorphic robots.

Anthropomorphic robots—also called humanoid robots—impose challenges on a wide multidisciplinary scientific community from mechanical and electronic engineering to computer, cognitive and social sciences. The ongoing efforts to address these challenges are (at least) twofold motivated. On the one hand, the technological innovations are driven to cope with the demand for service robotics including disaster relief, health care, elderly nursing care, household assistance and entertainment. On the other hand, there is an intrinsic human desire and need to understand the physiological and cognitive human constitution. These facts are driving diverse scientific endeavors to develop better comprehension—and consequentially a more accurate model—of the complex human structure. These efforts are concretely reflected in advances in biomechanics and artificial intelligence.

The pragmatic importance of anthropomorphic robots lies on their constitution and body structure. The morphologic composition of a humanoid robot allows the mapping between the human body and the physical structure of the robot. This fact leads to various advantages for humanoid robots compared to other robots or even automation systems. The main advantage of this mapping enables humanoid robots to handle the ubiquitous cursive equipment and infrastructure. This crucial aspect makes humanoid robots different to other systems allowing their seamless applica-

D. I. González Aguirre, *Visual Perception for Humanoid Robots*,
Cognitive Systems Monographs 38, https://doi.org/10.1007/978-3-319-97841-3_1

tion and integration into the society without modifying or requiring supplementary infrastructure.

Humanoid robots require autonomous and intelligent perception-action subsystems capable to accomplish complex coordinated skills for useful cooperative and interactive tasks. This covers wide engineering and research spectrum ranging from embodiment-dependent skills [1] like sensor fusion and motion control to pure-abstract skills such as symbolic reasoning and planning [2]. In particular, visual perception involves both types of skills for recognizing and interpreting the environment. The fundamental visual perception stages and questions are presented the following section.

1.1 Perceptual Fundamental Questions

The *multimodal*[1] skills of a humanoid robot should be organized in a technical cognitive architecture in order to execute an intelligent and knowledge-based perception-planning-action cycle [3]. In such an cognitive architecture, the role of perception is to transform sensor signals (*stimuli*) to internal representations (*percepts*) useful to generate directed actions and reactions to achieve tasks. It can be observed that—independently from the sensing modality—the perceptive skills for robot task execution have three interconnected functional stages:

- **Pre-execution alignment**: During this stage, the robot pose and state of the world have to be acquired and reflected upon the internal representation. The processes in charge of this representation alignment provide the initial and indispensable assertions and measurements for qualitative and quantitative task planning.
- **Execution trailing**: Perceptive skills are responsible for verification feedback during task execution. This implies prediction, detection and tracking of the conducting actions.
- **Post-execution alignment**: Once the task has been (at least partially) completed, perceptive skills determine the state of the world in order to reflect unplanned or unexpected changes upon the internal represented state of the world. This stage is tightly related to the initial pre-execution alignment. At this stage, a priori and a posteriori knowledge about the state of the world must be integrated.

Depending on the sensing modality, the perceptual capabilities of a robot are either concerned with the outer state of the world through exteroceptive sensing skills or are focused on the inner state of the robot by proprioceptive skills [4]. Due to its unobstructive nature and flexible applicability, visual sensing is the most dominant exteroceptive sensing modality for humanoid robots [5]. This can be partially accomplished by artificially emulating the human vision system through active stereo vision. In previously discussed functional stages, the active stereo visual perceptive skills have to reliably contribute to answer two tightly interrelated fundamental questions:

[1]The use of multiple heterogeneous input and output channels in a system.

- **What is it**? The answer to this question is the main objective of semantic perception [6]. This answer implies a linking path from raw sensor data to symbolic representations of the entities in the world. In order to bridge this signal-to-symbol gap, different processes have to be considered including sensing, attention and recognition. The goal of these processes is to generate a suitable representation of the entities in the world along with its efficient storage and management.
- **Where is it**? In addition to the recognition of objects, the visual perception system has to assert the position and orientation (*6D-pose*[2]) of the objects in the world in order to acquire the *scene-layout* of relevant objects and its relation to a given context. This spatial skills are essential for localization and interaction in terms of planning, navigation, grasping and manipulation. This means, active visual object recognition with pose estimation and visual self-localization.

1.2 Objectives

The principal research objective of this work is to discuss, understand and implement the essential environmental vision-model coupling for humanoid robots. Concretely, the coupling of model-based visual environmental object recognition with model-based visual global self-localization and its interrelated data handling are the ultimate goals to be achieved. These skills require novel methods to overcome the current sensing and matching limitations. The proposed algorithms and their efficient implementation in a platform independent software system has to enable humanoid robots to attain a bidirectional vision-model coupling. The developed methods apply active stereo vision and use geometric CAD models of real made-for-humans environments. The core applications are tasks involving self-localization and object recognition for grasping and manipulation. The scientific and technical goals of this work are:

- **Model-based visual environmental object recognition**: The humanoid robot has to dependably solve visual assertion queries. A visual assertion query is a request to recognize and estimate the 6D-pose of rigid environmental objects. The addressed environmental elements in this work expose nearly Lambertian[3] characteristics. For instance, door-handles, electric appliances, cupboards, etc.
- **Model-based visual global self-localization**: The humanoid robot has to reliably determine its 6D-pose relative to the reference coordinate system of the model. The estimated pose has to provide the description of its uncertainty distribution for its subsequent consideration in plans and actions.

---

[2]The pose of an object in 3D space is uniquely defined by a 3D vector location and three orientation angles.

[3]A lambertian surface is characterized by its isotropic luminance. This restriction is partially relaxed by metal and glass elements in the environment (see Chap. 5).

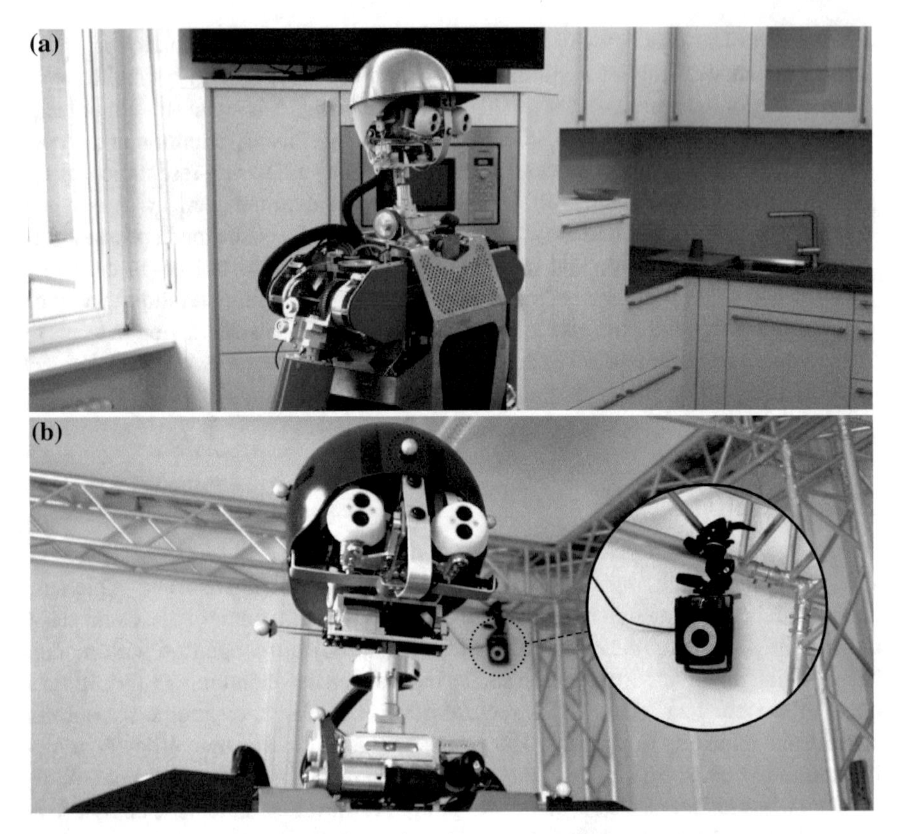

Fig. 1.1 The experimental platform. **a** The humanoid robot ARMAR-IIIa. **b** The humanoid robot ARMAR-IIIb and the capture system *Vicon Mx*®

1.3 Experimental Platform

The experimental application domain used in this study pertains real everyday made-for-humans environments such as the kitchen environment within the collaborative research center on humanoid robots of the SFB588.[4] The humanoid robots ARMAR-IIIa and ARMAR-IIIb [7] are the platforms where this work has been developed and experimentally evaluated. Figure 1.1 shows the humanoid robot ARMAR-IIIa in the so-called robot kitchen and the humanoid robot ARMAR-IIIb in the motion capture laboratory.

The humanoid robot ARMAR-IIIa has 43 degrees of freedom (DoFs) distributed on various subsystems; one active head, two arms with five-fingered hands, a torso and a holonomic platform [7]. The visual perception uses the active head of the humanoid robot exposing 7 DoF. Four DoFs are located within the neck as lower pitch, roll,

[4]The collaborative research center 588 *"Humanoid Robots - Learning and Cooperating Multimodal Robots"* founded by the *Deutsche Forschungs-Gemeinschaft* DFG.

yaw and upper pitch. The remaining DoFs are located in the eyes arranged into one common tilt for both eyes and two independent eye-pan actuators. Each eye consists of two CCD Bayer pattern cameras with a resolution of 640 x 480 pixels with and maximal frame rate of 60 FPS [8]. The humanoid robot ARMAR-IIIb and the motion capture system *Vicon Mx*® were used to provide the ground truth measurements for the development of the visual uncertainty model and its evaluation.

1.4 Contributions

The main contributions of this work are related to the robust and accurate visual manifold sensing needed for efficient processing of HDR images generated by the ARMAR systems. Second, a model-based visual object recognition method with high improved recognition robustness and accurate geometric resolution is proposed. Third, a model-based self-localization method is proposed for determination of the robot pose relative to areas of interest in the environment of the robots.

1.4.1 Robust and Accurate Visual Manifold Sensing

The focus of this work is related to the visual sensing process. Moreover, a novel methodology for image acquisition on humanoid robots was developed. This visual manifold sensing overcomes the limitations imposed by both the complex hardware of a humanoid robot and the environmental conditions in real applications. The proposed visual sensing method enables robots equipped with standard cameras[5] to acquire *quasi-continuous*[6] high-dynamic-range (HDR) images with high signal-to-noise ratio. This visual sensing method consists of two process:

- **Temporal-fusion**: The first step necessary to acquire the visual manifold is a method based on multiple image fusion for signal denoising and range enhancement [10]. The main advantage of the method is its pixel-wise stability and robustness against arbitrary multimodal distributions of the irradiance sensor signals. This contribution is supported by experimental evidence which clearly ratifies the improvement of the image and consequently a higher robustness of the visual recognition tasks.
- **Exposure-fusion**: The second step necessary to acquire the visual manifold manages the ubiquitous high-contrast image content. The severe rate-distortion quantization effects produced by both the single-auto exposure mechanism and the narrow resolution of the analog-to-digital converter of the camera limit the dynamic

[5]Off-the-shelf cameras providing range-quantized (8-bit), low-dynamic-range (LDR) based on color filter array CFA (Bayer pattern) sensors, such as the IEEE-1394a *DragonFly* cameras [9] in the humanoid robots ARMAR-IIIa and ARMAR-IIIb.

[6]Due to the raw images nature, the density property of real numbers is not hold.

range of the acquired images. Using low-dynamic-range (LDR) cameras, the proposed method [11] enables the acquisition of accurate HDR images to properly capture complex and heterogeneously lighted scenes. The use of HDR images with mobile robots remained unexplored despite of its evident advantages. The novelty and importance of the proposed HDR imaging method for robot perception relies on two elements:

- **Improved radiometric calibration**: An improved method for estimating the artifact-free radiometric calibration is proposed. Based on the proposed continuous, nonlinear and reciprocity consistent calibration model, the fusion of differently exposed images allows the synthesis of high-dynamic-range images with superior quality for image processing. This novel method significantly improves the HDR images accuracy for feature extraction.
- **Optimal exposure control**: The optimal selection of exposures is determined adaptively according to the radiometric calibration and the granularity of the exposure set of the camera(s). This selection minimizes the amount of exposures necessary to ensure high sampling-density reducing the visual manifold acquisition time.

1.4.2 Model-Based Visual Object Recognition

The appropriate visual manifold acquisition enables the consistent extraction of geometric-primitives of the scene with a high degree of independence from pervasive deteriorating photometric effects. State-of-the-art methods for geometric-primitive extraction are not capable to manage high-dynamic-range images without parameters (usually thresholds) which dependent on particular image content. Since these parameters can hardly be locally and automatically estimated, the robustness and applicability of the geometric methods for object recognition were limited. In order to address this problem, a novel non-iterative and parameter-free (namely, without image content dependent parameters) method for geometric saliency extraction and representation is presented [11]. The proposed method allows the precise recognition of complex objects in real world scenarios with varying illumination conditions. This is one major contribution of this work supported with experimental evidence. In this method, the integral solution of following scientific issues was achieved:

- **Improved recognition robustness**: Achieved by the proper extraction of structural saliences. This means, the selection of visual-features does not depend primarily on their saliency magnitude but mainly on their geometric constitution. This enables wide applicability through high sensitivity without performance trade-offs.
- **Accurate geometry from radiance saliency graphs**: The combination of the topological connectivity and the spatial arrangement of the structural image saliencies provides the propitious insight into the underlying geometric composition of the scene. A novel characterization of these combined cues profitably unveils and simultaneously segments the geometric-primitives such as line segments, cor-

ners and free-form curves. This approach offers various advantages compared to state-of-the-art methods. For instance, the proposed method exhibits no discretization in the local orientation of the geometric-primitives and (in contrast to other approaches) the radiance saliency graphs expose no degeneration when dealing with complex foregrounds or cluttered backgrounds.

1.4.3 Model-Based Visual Global Self-Localization

In model-based approaches, the visual recognition and model matching of these environmental elements contain enough information to estimate the 6D-pose of the robot. The extraction of this vision-model coupling requires a method which reliably creates a kinematic chain linking the visual perception to the world model representation. An important contribution of this work is a method for attaining this kinematic chain by simultaneously solving both critical issues, namely the data association (visual-features to shape-primitives) and pose estimation (shape-primitives to optimal pose assessment) through a novel multilateration concept. This is based on the intersection of probabilistic-and-geometric primitives called density spheres. The introduction of this new entity allows the simultaneous solution of the data association and pose optimization. This is achieved by means of the proposed closed-form solution for the robot location estimation. In these terms, the robot location is the point with maximal associated density resulting from the conjunctive combination of density spheres. In this formulation, the density profile of the spheres corresponds to the visual depth uncertainty distribution of the humanoid robot system. This uncertainty density profile is achieved in a sound novel manner within this work through supervised learning. Detailed description, mathematical derivation and algorithms are provided.

1.5 Outline

- **Chapter** 2 presents a perspective of the state-of-the-art on technology and methodology for visual perception for humanoid robots. The focus is placed on relevant work in robot vision for object recognition with 6D-pose estimation and vision-based 6D global self-localization. Especial emphasis is paid to visual perception capabilities in human-centered environments.
- **Chapter** 3 presents and discusses the aspects involved in the selection and development of the proposed geometrical and topological world model representation. The indexing structures and query mechanisms necessary for recognition (used during the visual assertion queries and visual global self-localization) are described. The detailed presentation of the transformations from standard CAD models (in

form of interchange files) to the proposed sensor-independent graph and geometric representation is provided. This is conveniently illustrated with examples of the experimental scenarios. The evaluation of the developed representation and querying is provided.

- **Chapter** 4 presents the proposed visual sensing method for humanoid robots. A consistent treatment of the irradiance signals for robust image acquisition and optimal dynamic-range expansion through image fusion is presented.
- **Chapter** 5 presents the environmental visual recognition method. A detailed presentation of the proposed method for extraction of 3D geometric-primitives is provided. The synergistic cue integration (homogeneity, edge and phase rim) for extraction of 2D geometric-primitives is presented. The consolidation of 3D geometric-primitives by calibrated stereo vision is introduced. The effectiveness and precision of the proposed method are experimentally evaluated with various natural and artificial illuminations on different environmental objects ranging from rather simple doors up to complex electric appliances.
- **Chapter** 6 introduces a novel uncertainty model supporting visual depth perception for humanoid robots. The supervised learning approach models the visual uncertainty distribution using ground truth data. The validation and discussion of the attained model are presented.
- **Chapter** 7 introduces the proposed method for matching visually perceived 3D geometric-primitives with the world model representation while optimally estimating the 6D-pose of the humanoid robot. The probabilistic and geometric closed-form solution for robot pose estimation is presented.
- **Chapter** 8 concludes the work with a summary, critical evaluation of the contributions and briefly describes the future work.

References

1. Baxter, P., and W. Browne. 2009. Perspectives on Robotic Embodiment from a Developmental Cognitive Architecture. In *International Conference on Adaptive and Intelligent Systems*, 3–8.
2. Brooks, R., and L. Stein. 1994. Building Brains for Bodies. *Autonomous Robots* 1: 7–25.
3. Patnaik, S. 2007. *Robot Cognition and Navigation*. Berlin: Springer. ISBN 978-3-540-68916-4.
4. Sinapov, J., T. Bergquist, C. Schenck, U. Ohiri, S. Griffith, and A. Stoytchev. 2011. Interactive Object Recognition using Proprioceptive and Auditory Feedback. *International Journal of Robotics Research* 30(10): 1250–1262. ISBN 978-0387853482.
5. Bar-Cohen, A.M.Y., and D. Hanson. 2009. *The Coming Robot Revolution: Expectations and Fears about Emerging Intelligent Humanlike Machines*. Berlin: Springer. ISBN 978-0387853482.
6. Markovich, S. 2002. Amodal Completion in Visual Perception. In *Visual Mathematics*. vol. 4, ed. Slavik Jablan. ISBN 978-3-642-34273-8.
7. Asfour, T., K. Regenstein, P. Azad, J. Schröder, A. Bierbaum, N. Vahrenkamp, and R. Dillmann. 2006. ARMAR-III: An Integrated Humanoid Platform for Sensory-Motor Control. In *IEEE-RAS International Conference on Humanoid Robots*, 169–175.

8. Asfour, T., K. Welke, P. Azad, A. Ude, and R. Dillmann. 2008. The Karlsruhe Humanoid Head. In *IEEE-RAS International Conference on Humanoid Robots*, 447–453.
9. PTGrey. 2008. *Dragonfly Technical Reference Manual*, 5-8 2008.
10. Gonzalez-Aguirre, D., T. Asfour, and R. Dillmann. 2011. Robust Image Acquisition for Vision-Model Coupling by Humanoid Robots. In *IAPR-Conference on Machine Vision Applications*, 557–561.
11. Gonzalez-Aguirre, D., T. Asfour, and R. Dillmann. 2010. Eccentricity Edge-Graphs from HDR Images for Object Recognition by Humanoid Robots. In *IEEE-RAS International Conference on Humanoid Robots*, 144 –151.

Chapter 2
State-of-the-Art

In this chapter, a perspective of the level of knowledge and development achieved on visual perception for humanoid robots is presented. The focus is placed on robot vision for object recognition with 6D-pose estimation and vision-based 6D global self-localization. A compact and representative selection of the contributions to visual recognition and localization for humanoids robots is presented in order to examine the achievements and limitations of previous work. This examination also clarifies various particular aspects which differentiate the visual perception for humanoid robots from other non-anthropomorphic systems. Subsequently, the structural and conceptual foundation of this work, namely the proposed arrangement of spaces and information flows for the environmental visual perception in humanoid robots is outlined. Notice that in order to provide a comprehensive insight into the context and contributions of this work, each of the subsequent chapters presents additional work on particular methods. In this manner, this chapter allows a global perspective on the state-of-the-art meanwhile the locally focused and precise discussions on particular methods are provided in the corresponding chapters.

2.1 Humanoid Visual Perception

The research on visual perception for humanoid robots is a special domain in robot vision. In the 70 and 80s, there were various remarkable pioneers working on this particular domain, namely the first anthropomorphic robots and the earliest contributions to vision-based environmental perception.

The very first humanoid robot *WABOT-1* (see Fig. 2.1a) was developed at Waseda university in Tokyo by I. Kato [1]. WABOT-1 was equipped with artificial eyes according to [2] as:

D. I. González Aguirre, *Visual Perception for Humanoid Robots*,
Cognitive Systems Monographs 38, https://doi.org/10.1007/978-3-319-97841-3_2

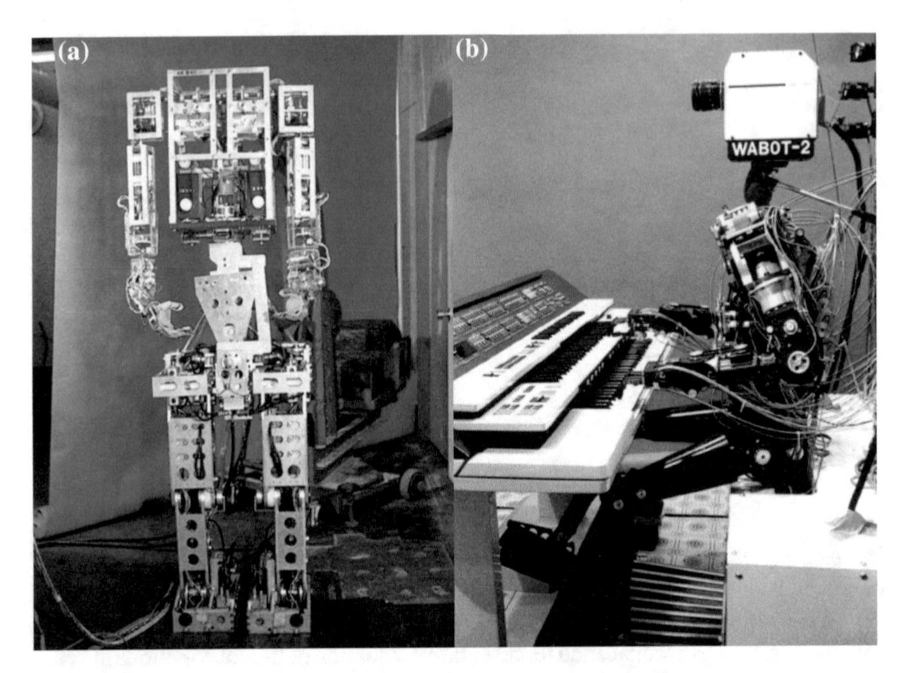

Fig. 2.1 **a** The first humanoid robot WABOT-1 [1]. **b** The musician specialist robot WABOT-2. Both images were extracted from [2]. [©1973–1987, Humanoid Robotics Institute, Waseda University Tokyo]

[sic]*"The WABOT-1 was able to communicate with a person in Japanese and to measure distances and directions to the objects using external receptors, artificial ears and eyes, and an artificial mouth".*

Despite having a visual system, the principal goals of WABOT-1 were focused on limb control, stabilization and system integration. Later, the successor humanoid robot *WABOT-2* [3] (see Fig. 2.1b) exposed very specific but remarkable complex skills. According to [2]:

[sic]*"The robot musician WABOT-2 can converse with a person, read a normal musical score with its eye and play tunes of average difficulty on an electronic organ".*

The capacity to read sheet music using images captured with a low resolution television camera was an impressive and certainly one of the very first vision-based object recognition and interpretation system in humanoid robots. In the following years, other similar projects emerged adopting similar goals and approaches. For example, the humanoid robot *Saika* [4] and the *WABIAN* (Waseda bipedal humanoid robot) [5]. Among these studies, an outstanding project in humanoid robot vision appeared at the end of the 90s in the MIT (Massachusetts Institute of Technology). The *Cog* humanoid robot [6] was developed with focus on developmental structure and integration of multiple sensors and motors for social interaction (see Fig. 2.2).

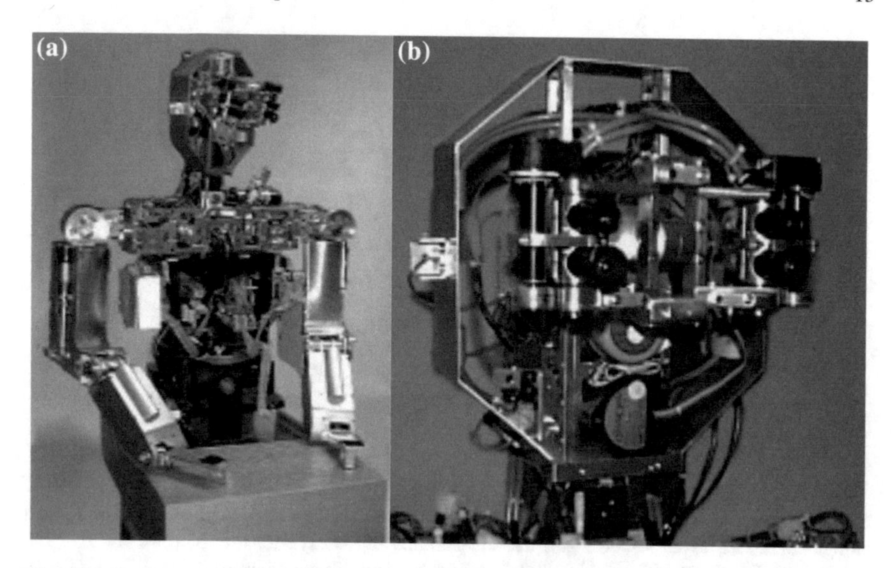

Fig. 2.2 The humanoid robot Cog. **a** The upper-torso with twenty-one DoFs distributed on the head and two arms. **b** The visual system included four NTSC cameras, two per eye (see [7]). The vergence was possible due to the pan rotation per eye and a coupled tilt axis. Extracted from [6]. [©1999- Springer]

In this upper-torso humanoid robot with 21 DoFs, various systems were integrated including visual, auditory, vestibular, kinesthetic and tactile sensors. The visual research was conducted in areas concerning the understanding of human cognition [8], particularly in smooth-pursuit for tracking, eye movements (in terms of saccades), binocular vergence and ocular reflexes just to mention a few.

In general, early vision, social behavior and human-robot interaction by expressive feedback were the goals of the Cog project. In subsequent years, the visual skills for manipulation and scene analysis were also investigated but only in a simplistic way in order to provide input for research on biological-like processes. Other research groups in academia and a few enterprises (Hitachi in 1985 with the WHL-11 biped robot, Honda in 1986 with the experimental model E0, Sony in 2001 with the Sony QRIO *"Quest for cuRIOsity"*, Fujitsu in 2001 with the miniature humanoid HOAP-1, etc.) all over the world remain interested in the research and application of humanoid robots not only as a platform for cognition research but also for education, entertainment, service and assistance. In order to make humanoid robots capable of such scenarios, there were (and many still are) challenges to be solved in terms of embodiment aptitudes and intelligence. From these early stages, there were two distinguishable research directions coping these challenges: On the one hand, the researcher focused on motion control, stabilization, walking, stair climbing and other dancing-like complex behaviors. This essential research direction is the one concerned with the intelligence necessary to manage the body of the humanoid. This sequential development in a noteworthy enterprise is shown in Fig. 2.3. On the other

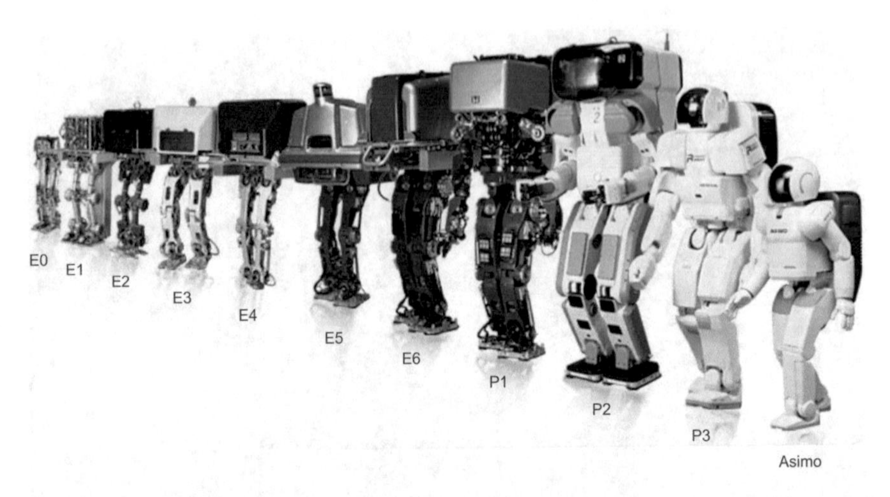

Fig. 2.3 The *Honda* robot evolution, from biped walking experimental machines to full integrated humanoid robots. From left to right; The first Honda biped experimental platform E0 in 1986 followed by E1, E2 and E3 (1987–1991). Later on, the experimental platforms E4, E5, E6 (1991–1993) evolved in terms of walking speed and terrain capabilities. Subsequently, the first Honda humanoid research platform was the P1 (175 kg @ 1,915 mm) with upper limbs and torso powered by external sources. Afterwards, the P2 (210 kg @ 1,820 mm) was the first platform capable of walking up and down stairs as well as pushing shopping-cars and other task with external powered sources. The P3 (130 kg @ 1,600 mm) was the first platform to have fingered hands and suitable weight for real applications in human-centered environments. Finally, the humanoid robot *Asimo* (2000–2011) is not only an improved version of the platform P3 in terms of weight and power requirements. Asimo also exposes autonomous capabilities including recognition of humans and objects as well as prediction and planning of complex trajectories. [©1987–2012 Honda Motor Co. Ltd.]

hand, there is an ongoing research focusing on the indispensable intelligence for recognizing and manipulating objects in order to realize general and complex tasks. Since the first research direction is a fundamental one, there is a tendency to assume that this must be completely achieved in order to move towards the second direction. However, both research directions are tightly complementary (see more on this topic in [9]). In the case of humanoid robots with static hips or wheeled platforms, the research on the second direction could omit or simplify various aspects of the stabilization, walking and whole body balancing. Consequently, by employing such physical embodiments, it is possible to concentrate on the problems of recognition and manipulation. A notable example of these endeavors at early stages was the action selection in reaching and grasping objects based on vision in [10]. In this work, the authors (M. Inaba and many others in his research group) used the so-called *"remote brain robot"* approach which enables external computations outside the robot. The humanoid robot with 35 DoFs could stand up, reach and grasp objects coordinating legs and arms. In these experiments, the authors applied model-based vision to recognize static polyhedral objects [11].

Another early example group was the visual guided homing for humanoid service robots in [12]. This was one of the first approaches of visual self-perception (particularly for servoing) on humanoid robots. In that contribution, the humanoid service robot HARO-1 used dense reconstruction methods from stereo in order to conduct geometric and kinematic analyses of its own arms. Despite of its limitations in accuracy, camera calibration and reconstruction quality, the experiments showed that the visual modality provides flexibility and adaptability for humanoid robots. Furthermore, there are various significant contributions [13, 14] using the humanoid robot platform iCub [15, 16]. These recent contributions focus either on single object manipulation or developmental learning. The parallel evolution in terms of embodiment capabilities, computational power and robot vision methods gradually enabled humanoid robots to cope with more complex and realistic challenges. Notable contributions [17, 18] illustrate these advances using the experimental platform of this work ARMAR-IIIa,b (see Sect. 1.3).

2.1.1 Visual Self-Localization

The following survey presents contributions to humanoid vision for self-localization and object recognition methods with 6D-pose estimation. The selection of the survey is based on impact within the robot vision community and novelty of their approaches.

Model- and Stereo-Based Self-Localization

The first notable work on visual-based self-localization on humanoid robots is the contribution of S. Thompson et al. in [19] (see Fig. 2.4). Similar to most of the early works in visual perception for humanoid robots, the authors exploited dense stereo reconstruction. The reconstruction was considered superior to local visual-features such as point features (for instance, the scale invariant feature transform, SIFT in [20]) due to the lower data association complexity involved during matching and the necessary information for footstep planning. The visual localization conducted in [19] is actually a 3D reduction (two DoF for position and one DoF for orientation) of the 6D problem by assuming a flat ground. The arguments supporting this strong reductionism were the inherent noise of dense stereo vision, high computational cost of 6D search, the availability of robot odometry and possible dimensionality reduction of planar localization. The humanoid robot H7 (see Fig. 2.4a) was the experimental platform of the method [19]. This robot was equipped with the SVS (Small Vision System [21]) which provides low resolution (320×240 pixels) images. Thus, the minimal and maximal depth were limited to 1–3m.

Despite the strong assumptions and limited degrees of freedom, the authors were capable to apply state estimation based on sequential Monte Carlo filtering [23]. Specifically, the particle filter was used to determine the state vector S_t at time t

$$S_t := [x_t, y_t, \theta_t]^T. \tag{2.1}$$

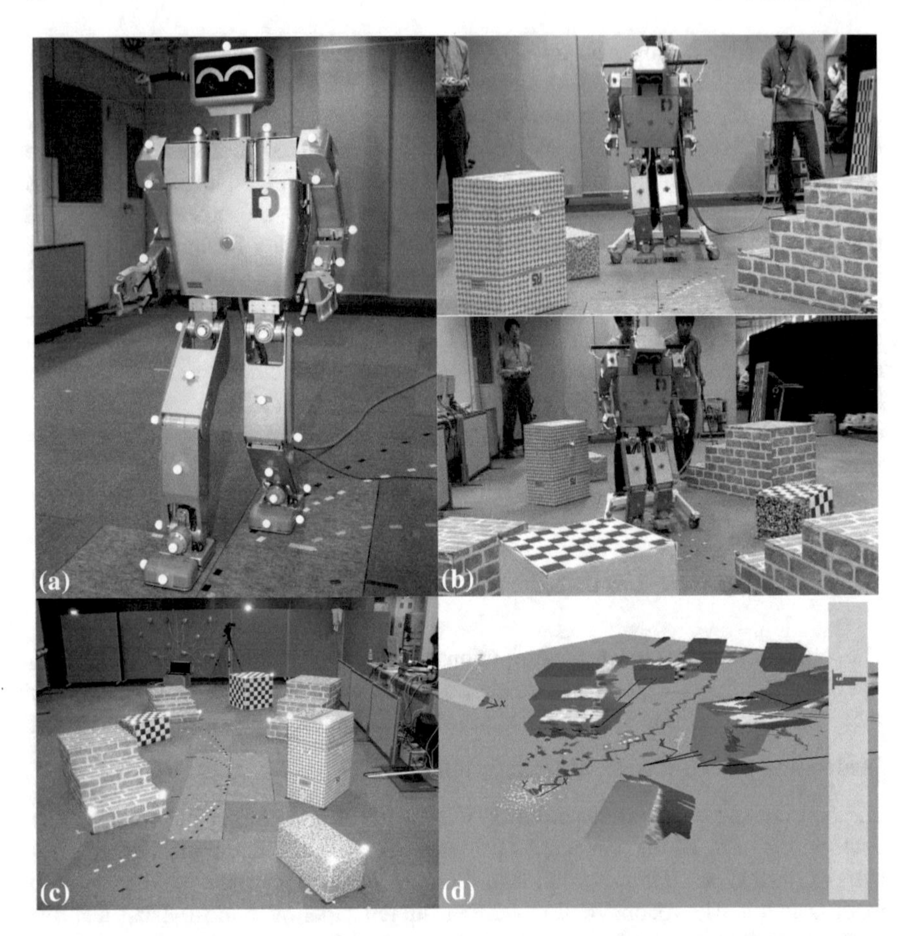

Fig. 2.4 The model- and vision-based self-localization of the humanoid robot H7 [22] by dense stereo reconstruction and particle filter. **a** The humanoid robot H7 has spherical markers located at strategically locations in order to determine its reference pose during the experimental evaluation. **b** Two video frames showing the walking path of the robot while visual self-localization takes place. **c** The experimental setup for visual self-localization. **d** The results of the experimental evaluation are visualized within the virtual environment by overlaying the sensor reconstruction and the ground truth walking paths attained with the markers. Extracted from [19]. [©2005 - IEEE]

The application of this method requires three models:

- **Environment model**: Made from known dimensions of the obstacles which were registered by markers (see Fig. 2.4c).
- **Sensor model**: Formally described in terms of the probability $P(v_t|s_t)$. It quantifies the likelihood of having the dense reconstruction profile v_t at the pose s_t. It is expressed as

$$P(v_t|s_t) = \prod_{i=1}^{N} P(d_i|s_t), \tag{2.2}$$

where d_i represents the distance from the dense reconstruction surface to the camera along the optical axes located at N discrete angular steps of the field of view. This probabilistic sensor model $P(d_i|s_t)$ determines the likelihood using the distance between the measured depth and expected depth from the virtual model z_i^d. This similarity was bounded by a Gaussian kernel as

$$P(d_i|s_t) = \frac{1}{\sqrt{2\pi\sigma^2}} \mathbf{exp}\left[\frac{(d_i - z_i^d)^2}{-2\sigma^2}\right], \tag{2.3}$$

where the standard deviation σ depends on the depth as

$$\sigma = \left(1 + \mathbf{exp}[-d_i * 0.001]\right)^{-1}. \tag{2.4}$$

This depth deviation incorporates the disparity resolution effects (see a detailed model of this aspect in Chap. 6).

- **Motion model**: This was done with the robot odometry using the kinematic model and joint encoders of the humanoid robot H7 without explicitly considering friction, sliding or other drifting effects.

The experimental evaluation of this approach was realized as illustrated in Fig. 2.4b. The results showed that the application of the particle filter in humanoid robots is a plausible and powerful technique to incorporate environmental knowledge while integrating sensor and motion commands in a noise tolerant manner. Despite the achieved results, the necessary texture in the environment, the computational complexity of the dense reconstruction (even with reduced precision) and the reduced nature of the environment imposed serious limitations for a general application of the approach. In summary, this method is a model- and vision-based dynamic localization which enables the tracking of the robot pose by means of Bayesian filtering. The global or initial localization needs to be known in advanced in order to initialize the hypothesis particles. The main contribution of the work is the use of particle filters as an adept mean to integrate uncertain observations with model-based world representations. The main disadvantage of the approach is the limited visual-feature representation due to the textured requirements. Another concern is the tight dependency on odometry. This is an important issue because, in general, the drift produced by sliding or friction effects limits the applicability of the proposed method (see quantitative analysis on this issue in the motivation of [24]).

Monocular-Based Self-localization

A significant contribution to real-time simultaneous localization and mapping SLAM[1] using monocular vision for humanoid robots was presented by Stasse et al. in [24].

[1] "*Simultaneous Localization And Mapping*", is the wide spread usage of the acronym http://openslam.org/. The first appearance of the acronym was at the "*Software Library for Appearance Modeling*" in the work of H. Murase et al. [25] http://www.cs.columbia.edu/CAVE/software/softlib/slam.php.

This approach applies a different strategy for the visual localization of humanoid robots [19]. This method for dynamic localization is an extension of the monocular visual SLAM of A. Davison et al. [26, 27] by integrating information from the pattern generator and various proprioception measurements such as odometry and a 3-axes gyroscope during the state estimation. In that work, the camera state vector $\hat{\mathbf{x}}$ is composed of 3D position vector $\mathbf{r}^W$, orientation quaternion $\mathbf{q}^{WR}$, velocity vector $\mathbf{v}^W$ and angular velocity vector ω^R. The visual-features used as landmarks are sparse point-features (SIFT [20]) with their associated image patches (11×11 pixels, see Fig. 2.5). In the initial work of Davison, the camera self-localization process required no odometry neither other source of information. The camera pose was modeled with perturbations with Gaussian distributions in translation $\mathbf{V}^W$ and Ω^R rotation velocities. This model assumptions enable the application of the Extended Kalman Filter (EKF) to estimate and track the camera pose. This process is formulated in discrete time Δt described as

$$f_v = \begin{pmatrix} \mathbf{r}^W_{new} \\ \mathbf{q}^{WR}_{new} \\ \mathbf{v}^W_{new} \\ \omega^R_{new} \end{pmatrix} = \begin{pmatrix} \mathbf{r}^W + (\mathbf{v}^W + \mathbf{V}^W)\Delta t \\ \mathbf{q}^W R \times \mathbf{q}((\omega^R + \Omega^R)\Delta t) \\ \mathbf{v}^W + \mathbf{V}^W \\ \omega^R + \Omega^R \end{pmatrix}. \tag{2.5}$$

Furthermore, this approach was extended by incorporating angular velocities measured by a 3-axes gyro sampled at a 200 Hz from the humanoid robot HRP-2. The formal inclusion of this information was solved by unifying the camera frame and the gyro frame in the EKF formulation of the processes (from Eq. 2.5). The experimental evaluation (see Fig. 2.6) with two elliptical trajectories showed the plausibility of the extended approach to enable the dynamic localization and mapping in humanoid robots in general environments. Although this visual SLAM implementation was capable to dynamically localize the humanoid robot, the method is capable to track only an initial state which is determined by a small set of manually annotated markers:

"*When only based on vision the single camera SLAM currently requires a small set of features* ***with pre-measured positions*** *to be specified to bootstrap tracking. In our experiments, a set of well-chosen salient features (a mixture of natural features and some* ***hand-placed targets*** *)* mostly on the wall in front of the robot's starting position ***were chosen by hand***. *It is relatively important that these* ***features are well-spread in the image at the start of motion*** *since they effectively seed the rest of the map, which will be* ***extrapolated from the initial robot motion*** *estimates they provide.*" Extracted from [24].

This extract from [24] states that strongly constrained setups are required for the system to work. It also expresses the limited nature of the map, namely a representation not useful for robot interaction, (see discussion and proposed world model representation in Chap. 3). In [24], the robot is capable of dynamically localize itself only after being globally localized per hand on salient markers. The authors discussed an alternative to reduce these hard constraints by means of robot odometry. In any case, the asserted position and orientation are only linked to the initial unknown pose

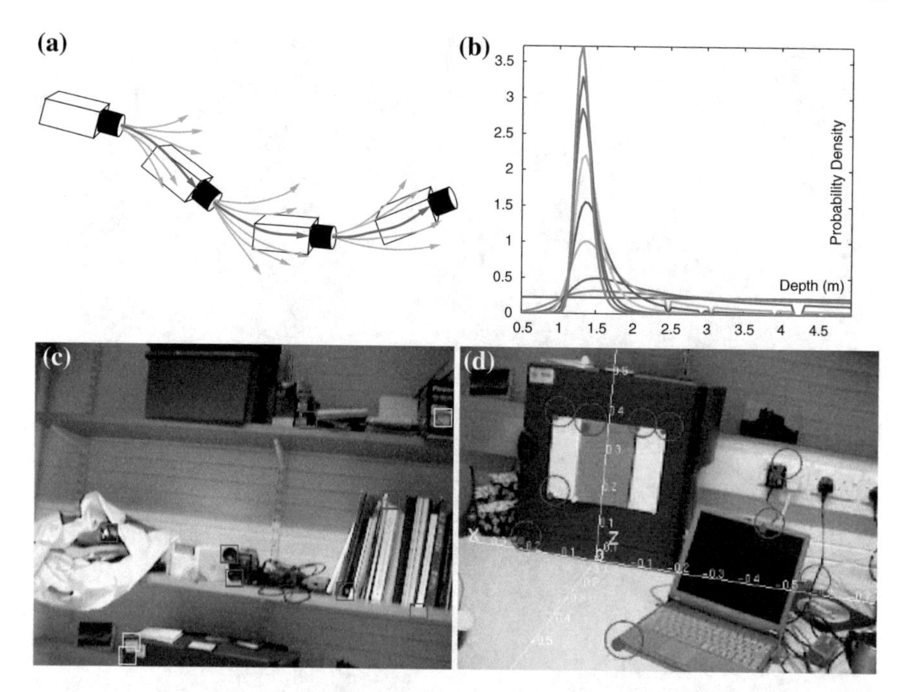

Fig. 2.5 The monocular visual SLAM [27] is the basis of the dynamic visual self-localization of the humanoid robot HRP2 [24]. **a** Visualization of the constant velocity model Eq. 2.5. **b** The feature probability distribution. **c** The feature points with image patches. **d** Regions where the visual-features are estimated. Extracted from [24]. [©2006 - IEEE]

of the robot. This problem cannot be tackled by similar schemas. This requires novel representation(s) and data association method(s) to optimally match visual-features while classifying objects in the environment.

In summary, the pure vision-based self-localization of humanoid robots is only partially solved. The well-established paradigm for vision-based dynamic localization is the result of intensive research in this field over the last two decades. In particular, the work of S. Thompson, O. Stasse and A. Davison clearly shows the critical gap between these paradigms for humanoid robot vision, namely the vision-based global localization with application specific environmental representations. The global localization implies two important elements: (i) A world model representation to globally reference the attained poses and recognized objects. (ii) A scalable data association method to match the uncertain visual-features with the shape-primitives of the object models in the world representation.

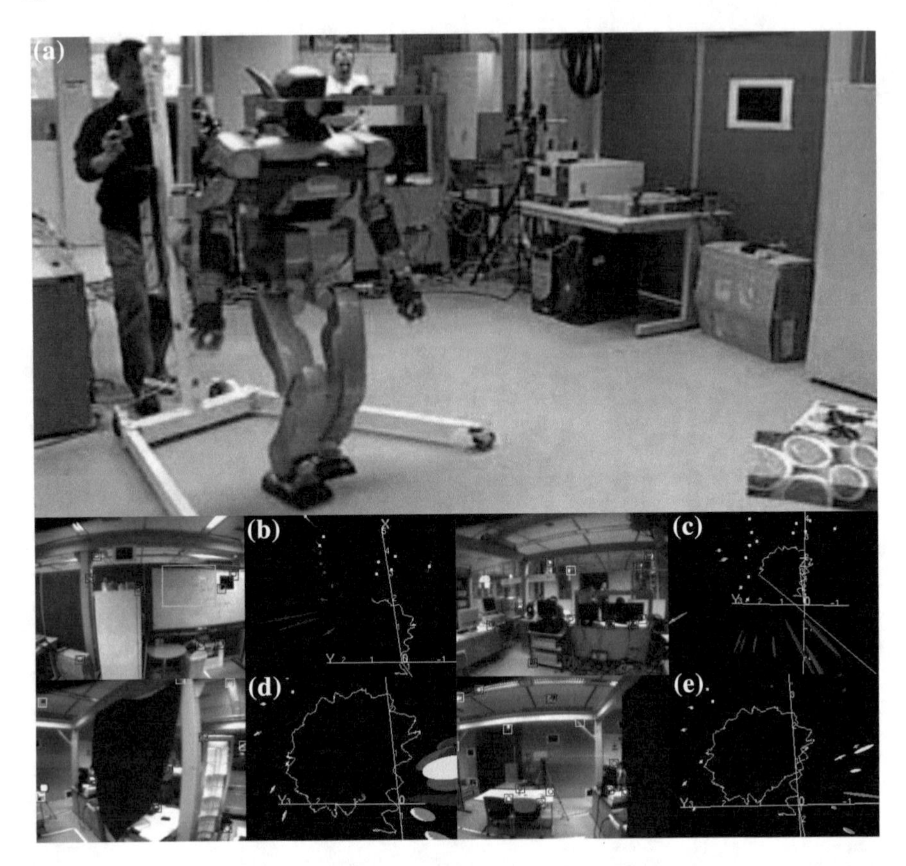

Fig. 2.6 Experiments on visual SLAM in the humanoid robot HRP-2 using pattern generation, monocular vision and gyroscope information in a circular trajectory with 0.75 m radius. **a** The humanoid robot HRP-2 walks autonomously while simultaneously localizing and creating a scattered map. **b** The visual-features (on the left image) and estimated state (in the 3D visualization window) are shown at the first turn of the walking trajectory. **c** Notice the rather large uncertainty distribution ellipses created while finding visual-features on the back wall. **d** The maximal uncertainty is reached shortly before the loop closure. **e** Afterwards, the loop closure is completed. The drift is corrected and the sparse map is adjusted. Extracted from [24]. [©2006 - IEEE]

2.1.2 Environmental Visual Object Recognition

The application of humanoid robots in everyday scenarios requires high level of intelligence for perception, planning and action coordination to efficiently realize given tasks. There are various key elements to achieve such complex robot behaviors. Some of these key elements have been partially shown in the work of K. Okada et al. particularly the contributions [28–30]. The conceptual approach of this work provides the robot with a full-fledged world model representation containing necessary information to allow the robot to accomplish complex assignments. The implemented representation accelerates the developmental process compared to other conceptual approaches such as those in artificial intelligence and neuroscience as discussed in [31]. Since this knowledge-based representation supports various complex problems (for example grasp planning or focus of attention) for task execution, the sensor-based adaptation to partially modeled, unknown or dynamic situations was the contribution of K. Okada et al. Obviously, handcrafted representations are not suitable for general and autonomous application of humanoid robots. However, the integration of sensor-based verification methods in a flexible manner was an important step toward autonomous humanoid robots. The key elements to be integrated in a vision-based behavior verification system for humanoids for daily tasks are:

- **Unified representation**: The first key is an integrated representation for motion planning [32] and sensory verification including vision and force sensor. This integrated representation has been adapted and extended by other researches with interesting results, for example, fusing vision with force in [33, 34].
- **Knowledge models**: The second key element is the incorporation of knowledge representation models of force events, visual attention and manipulation strategies (see examples of diverse information sources in Fig. 2.7). The information added to the world representation and the derived capabilities are directly coordinated with a behavior management. The particular information used to coordinate these tasks consists of the following elements:
  - **Information about location**: A 6D frame where the humanoid robot has to be located in order to visually capture environmental and non-attached objects (see the coordinate frame in Fig. 2.7a).
  - **Grasp information**: Defined either as a 6D frame or a subspace relative to the geometric description of the objects. This is an extended object-centered representation which notably reduces the search space for grasping. It also fixes the target to *"look at"* while an action is being conducted (see examples at the oven, plate and kettle on the right side of Fig. 2.7a).
  - **Manipulation information**: Appears in two different forms depending on the object and the desired behavior: (i) As a single 6D frame where the robot has to apply contact in order to realize a target transformation. (ii) As a sequences of 6D-poses of the object in order to realize a task, For instance, vacuuming the carpet or puring by manipulating the tip of a kettle (see Fig. 2.7b).
- **Behavior management**: Consists of a planner-based motion component coupled with a sensory verification system. The sensory verification system is capable to determine which sensing modality will provide the most important information during each phase of the task execution. This means, the system is able to generate and verify the motion plans not only by detecting and updating locations but also by predicting and recognizing events such as contacts with the furniture and other forms of complex awareness, for instance, visual detection of water flow.

The visual perception methodology of the approach has been presented in [29]. The core of this methodology is a multi-cue 3D object recognition process. It incorporates

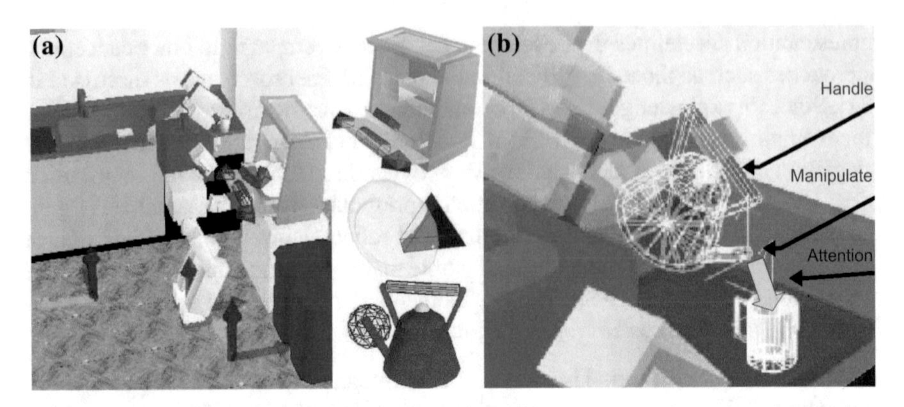

Fig. 2.7 The enhanced world model representation for complex tasks by the humanoid robot HRP-2. **a** The handcrafted knowledge of spot on the floor, handle represented by triangular prisms and cylinders attached to the objects. The attention subspace is shown with an example kettle (depicted with a sphere at the tip). **b** The *"sensory navigation"* is illustrated with behavior verification. Extracted from [28]. [©2006 - IEEE]

CAD models, visual-features and color histograms. This visual perception exploits the world model representation in two ways (Fig. 2.8):

- **Quasi-perceptual imagery**: An internal long-term and short-term memory by means of *mental imagery*[2] in order to attain visual-features from hypothetical configurations (both from robot and objects). These hypotheses were used for the similarity measurement during state estimation.
- **Feature selection**: A priority selecting source of salient visual-features, concretely the selection of robust visual-features from CAD models.

The strategy used by the authors was an extension of the ideas previously presented in [19]. In particular, the integration of models into the visual perception by means of particle filters was revisited. Nevertheless, the novelty of this approach was the inclusion of three different cues merging geometric and appearance information:

- **3D-Shape cue**: The first cue, the so-called *shape* information is a hybrid concept exploiting the distance from the camera to the object(s) in the field of view of the humanoid robot. This cue also contains information of the object geometry in terms of surface subdivision. The formal representation is the probability $P_{shape}(z|x)$, where z represents visual and model information given the full body-pose and object-pose x. This likelihood is obtained as follows:
  - First, the selection of a set of salient points is done with the help of a structural tensor analysis which is a concept found in various visual-features such as Harris [36, 37] or Shi-Tomasi features [38].

[2] A quasi-perceptual experience, it resembles a visual experience, but occurs in the absence of the external stimuli (see [35]).

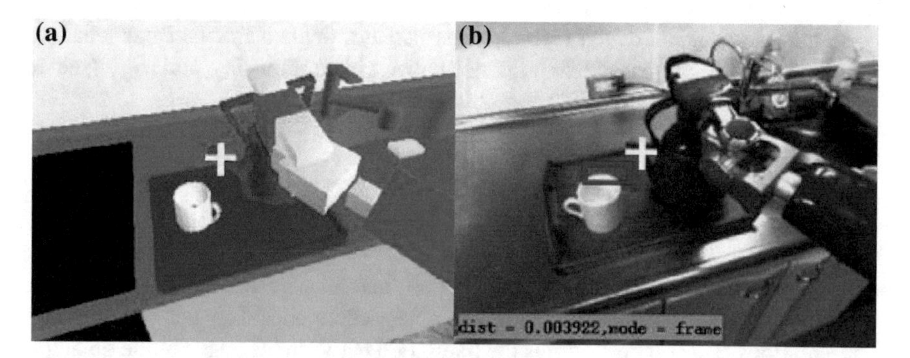

Fig. 2.8 Experimental results extracted from the approach in [28] for visual verification during task execution in humanoid robots. **a** The model-based unified representation for motion planning and visual recognition. **b** The input image of the humanoid robot with the marked attention zones. [©2006 - IEEE]

  - The extracted point feature sets from the left and right images of the stereo rig were matched by means of correlation. The camera calibration is used in order to estimate the 3D location of the point features. This 3D point set P is coupled with the CAD model of the object(s) by the inclusion of a depth threshold and the selection of visible reference faces of the objects $F_{ref}^{visible}$.
  - These 3D point sets and visible faces of the object(s) are integrated by a distance metric between salient points and points on the surfaces. This was denoted by the function D_{point}. This concept is integrated using a Gaussian kernel as

$$P_{shape}(z|x) = \mathbf{exp}\left[\frac{\frac{1}{|P|}\sum_{p\in P} D_{point}(p, F_{ref}^{visible})^2}{2\sigma_{shape}^2}\right], \tag{2.6}$$

  where the standard deviation σ_{shape} was user defined. In summary, this cue introduces information from appearance salient points by estimating the similarity to those points found on the scene object(s) (see Fig. 2.9).

- **HSV-Color histogram cue**: For each composed hypothesis containing object- and robot-pose, a color-based probabilistic metric is estimated as follows:
  - The rectangular region where the object is completely visible on the image plane is used to extract a color histogram h_{B_x} represented in HSV color space.
  - The same operation is then executed using virtual cameras and object(s). The resulting histogram is called references histogram and is denoted as $h_{B_{ref}}$.
  - Finally, the estimation of the similarity between both histograms is done based on the *Bhattacharyya distance* (see concept in [39]) plugged into a Gaussian kernel

$$P_{color}(z|x) = \mathbf{exp}\left[\frac{B(h_{B_x}, h_{B_{ref}})^2}{2\sigma_{color}^2}\right], \tag{2.7}$$

where the standard deviation σ_{color} was also user defined and the Bhattacharyya distance B is expressed using the normalized histograms h_{B_x} and $h_{B_{ref}}$ (see an example of this similarity in Fig. 2.9c) as

$$B(h_1, h_2) = \left(1 - \sum_{b=1}^{Nb} \sqrt{h_{b,1} \cdot h_{b,2}}\right)^{\frac{1}{2}}. \qquad (2.8)$$

- **2D-Straight edges cue**: A cue specially designed for objects with texture or color saliency was proposed using a mixture of object-centered and viewer-centered representations. This process attains the likelihood between the edges of the object(s) in the scene and those in the configuration hypothesis. This likelihood is denoted as $P_{edge}(z|x)$ and its computation consists of three phases:
  - The image edges are extracted by the filter in [40]. From those edge pixels, a subdivision approach is performed to extract the straight edges as short line segments denoted by E^{2D}.
  - The model edges of the CAD object are projected to the camera plane according to the hypothetic configuration (6D-pose of the object and configuration of the robot). The resulting edges are called reference line segments E^{2D}_{ref}.
  - In order to determine the similarity between both sets, a fixed length subdivision of the reference edge line segments E^{2D}_{ref} is done, namely $e^1_{ref}, \ldots, e^M_{ref}$. Later, these subsegments are matched against the real image edges E^{2D} by a thresholded function $D(E^{2D}, E^{2D}_{ref})$, namely a normalized angular and euclidean distance between segments. The similarity distance is plugged into a Gaussian kernel as

$$P_{edges}(z|x) = \mathbf{exp}\left[\frac{D(E^{2D}, E^{2D}_{ref})^2}{2\sigma^2_{edges}}\right], \qquad (2.9)$$

  where the standard deviation σ_{edges} is (in the same manner as in the shape and color cues) user defined.

2.2 Discussion

In the last decades, the research on environmental robot vision has focused mainly on the creation of maps for localization and navigation. The duality of the problem faced within this field has been expressed in the acronym SLAM (Simultaneous Localization And Mapping). There are remarkable advances in this field, particularly in humanoid robots or more general platforms using vision [41–43]. However, the developed methods and proposed representations are not suitable for interaction in terms of grasping and manipulation with humanoid robots because of the following reasons.

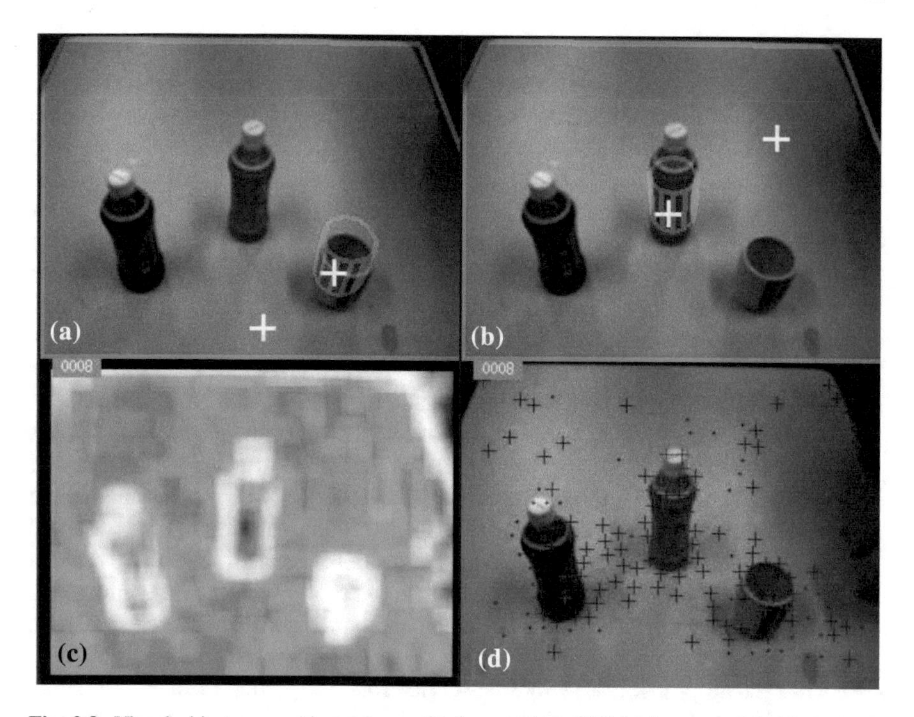

Fig. 2.9 Visual object recognition using multiple cues from [29]. **a** A successful object match. However, the pose still needs to be refined. **b** A miss match trial. The bottle was recognized as a glass. **c** The histogram cue similarity map $P_{color}(z|x)$ (Eq. 2.7). **d** The salient 3D feature points are illustrated for the same scene. [©2007 - IEEE]

- **Sparse representations**: Due to the data association complexity [42, 44], the commonly used visual landmarks for robot state estimation are only partially significant. This occurs systematically in the visual SLAM community through the application of the common paradigm, namely extracting, storing and matching of point and appearance features such as SIFT [20] and SURF [45] (see approaches using this paradigm in wheeled robots [26] and the previously discussed biped humanoid robot in [24]). Point features (landmarks) used in these methods provide scattered and viewpoint-dependent information, limiting the visual awareness to those combined feature regions and viewpoints. There is a recent tendency to integrate more significant landmarks like line segments [46] and other higher-order structures as presented in [41, 43]. The early advances in this direction are very promising for comprehensive visual perception.
- **Semantic enrichment**: In relevant contributions to visual localization (see [24, 27]), the assessed poses are only linked to the unknown initial pose (see criticism in Sect. 2.1.1). This crucial fact strongly restrains these approaches specially in the case of humanoid robots because these kind of robots require plenary environmental information to consistently interact with their environment. This limitation arises from the lack of semantic association, specifically, the semantic connec-

tion between visual-features and world model representation. Note that the visual semantic association is more than an unidirectional connection between the absolute 6D-pose of the robot and the world coordinate system. This visual association is the first missing link necessary to establish a generalized bidirectional visual linkage delineated in Sect. 2.4. Solving these issues requires a novel data association paradigm using semantic endowed representations.

- **Metric enrichment**: According to recent literature, the uncertainty distribution and precision of the pure visual SLAM methods are attained only in simulation or are not properly compared to reliable ground truth data. At the present time, there are very few works in the field of humanoid robots (see [19]) which compare their estimated poses against ground truth systems. Even in these approaches, the authors present no quantitative results and their only qualitative comparisons are visualization of the 2D navigation path of the robot. In the majority of the visual SLAM methods, results are generally qualitatively evaluated in terms of compelling subjective impressions on image overlay or other similar augmented reality applications. This occurs due to the lack of metric representations coupled with appropriate means to attain ground truth 6D-poses of the robot. This is a key aspect to be addressed to empower humanoid robots with the necessary skills to precisely recognize environmental elements for self-localization, grasping and manipulation.
- **Sensing paradigm**: Recent works in service robotics partially achieve semantic and metric enrichment (see [47–51]). However, they apply range sensors such as lasers [49] or structured light projection [50] in different segments of the spectrum (see a recent approach with the humanoid robot HRP-2 in [52]). Nevertheless, these solutions are restrictive due to time performance and because of the application of active-sensors makes it impossible for multiple robots to operate in the same space. Moreover, the sensing range that these systems can capture are reduced to short-range indoors applications. However, in the case of medium- to long-range applications, the necessary sensing devices are rather large, heavy and require high power consumption which is another limitation in light mobile robots. Based on these restrictions, these approaches are still inadequate for humanoid robots. Therefore, the passive vision paradigm used in this work is still the most appropriated sensing paradigm for humanoid robots.

The enhanced world model representation discussed in Sect. 2.1.2 shows its proficiency for complex task. However, the key elements (gasping planning, focus of attention, visual-feature labeling, etc.) are partially solved by hand annotations. Hence, its application is limited for general scenarios where the humanoid robots have to interact with previously non-annotated environments. Nevertheless, this approach revealed the importance of effective and efficient sensor-based prediction, verification and adaptation. There is a wide variety of approaches based on model-based visual object recognition using multiple cues such as edge, rim and region segmentation (for a detailed presentation see the surveys [53–56]). In Fig. 2.10a, simple and salient structures in minimalistic CAD models were the dominant type of object(s) in real-time tracking and grasp planning [57]. Approaches using single color free

Fig. 2.10 Assumptions in visual object recognition for humanoid robots

from objects have been successfully achieved in humanoid robots (see Fig. 2.10b, [58]).

This reduced the flexibility of the method due to the necessary overhead, extensive representation size, dependency on full image containment and ideal segmentation. The use of high contrast background (colors and materials) in controlled lighting setups produce (prominent saliency) shortcuts for the simplification of the visual perception (see Fig. 2.10c). In addition, the a priori scene layout (implicitly or explicitly implemented) in the methods restricts their application in general scenarios (see [30]). Moreover, various approaches which are robust to occlusion and lighting conditions have been proposed in humanoid robots [59]. However, the object complexity, the texture assumption and the appearance dependency are limiting factors for the application in the environmental vision (see Fig. 2.10d). Most of these methods are unsuitable for humanoid robots because they hold assumptions preventing their application for environmental perception involving self-localization, grasping and manipulation. Assumptions which make these approaches difficult to handle environmental visual perception are:

- **Containment**: Most of the approaches focus on simple structured objects which are easily to segment. These objects are significantly smaller than the 3D field of view of the humanoid robot [59]. Thus, the application domain consist of objects fully contained in the images even at manipulation distances.
- **Prominent saliency**: The domain of application is reduced to objects which present striking characteristics either in texture, color, contrast or shape [60, 61].
- **Reduced complexity**: Most of the existing works reduces the complexity of the objects and scenarios [29]. There are few methods which overcome this restrictions [58]. However, these approaches require extensive sensor- and instance-dependent training to produce efficient matching representations.

- **A priori information**: Especially in humanoid robots, the visual recognition algorithms are often reduced to partial detection. This occurs when the methods are explicitly or implicitly provided with background information about the scene. Usually, this information indicates which objects are in the field of view while providing hints about their spatial distribution, for instance, *"on the top of a table"* [29] or *"tray"* [30]. This a priori information makes the methods unsuitable for general environmental visual model coupling.
- **Controlled environment**: Even the relative low complexity of structured environments such as offices or houses are still difficult to handle by the visual perception system of humanoid robots. This occurs because existing approaches do not consider sufficiently the importance of the visual manifold acquisition and focus often only on feature extraction and matching mechanisms. This results in a low quality signals flow from the physical space into the visual space. The acquired low quality visual manifold destabilizes or prevents the successful application of almost all feature extraction methods. For instance, scenarios with slightly deficient illumination are still difficult to handle. The same situation occurs when an ubiquitous wide-intra-scene radiance range is found. For example, the direct light source like a lamp or the back-lighting of a window in the background.

These four issues cause serious problems which have to be solved in order to equip humanoid robots with a general 3D model-based object recognition system for self-localization, grasping and manipulation.

2.3 Environmental Visual Perception Skills

In the literature, self-localization is divided into two categories, *fine* (also called dynamic) and *global* localization (see [62]). The first category deals with the continuous state (dynamic-pose) of the robot and the second category is addressed in this work. It determines the 6D-pose of the robot in the world coordinate system (see Chap. 3).

Numerous approaches and robust solutions yield to the use of visual SLAM to solve the vision-based dynamic localization problem. However, less attention has been given to global localization with task-specific maps, namely localization with semantic and functional knowledge necessary for humanoid robot applications. This situation occurs due to: (i) The lack of mature sensor-based methods for the automatic generation of task-specific world model representations. (ii) The absence of robust, efficient and scalable methods for visual entity association.

In this work, the focus is placed on the solution of the second aspect. Specifically, emphasis is placed on the association between visual-features and shape-primitives in order to achieve a robust and precise global localization. The problems to be solved are:

- **Visual manifold acquisition**: How to attain an optimal visual transducer connecting the visual space with the internal representations of objects and places.

- **Data association**: How to efficiently associate visually acquired geometric structures with a world model representation.
- **Perceptual uncertainty**: How to model and manage the combined perceptual and actuator uncertainty arising from the visual recognition methods and actuator deviations.
- **Pose estimation**: How to optimally integrate the information provided by multiple perceptual structures according to their uncertainty in order to obtain a reliable and 6D-pose estimation.

In this work, prominent object saliency and controlled circumstances in the environment are not assumed. On the contrary, the materials (surface properties) and illumination of the experimental setups (see Sect. 1.3) have been intentionally chosen without simplification to achieve a general visual perception. Furthermore, since the humanoid robot has to be able to visually explore its surroundings, the visual containment is not a plausible assumption for the objective of this work. In other words, the only a priori information available for this visual perception system is *"the humanoid robot has to be within the physical space modeled in the world model representation"*. Regarding the complexity of the environmental objects, the scope is set to those objects typically found in household scenarios or office environments, such as electric appliances, furniture, walls, windows, etc. (see Chap. 5). In order to be useful for vast domains, the particular world model and the humanoid robot embodiment used in the system development and experimental evaluation must not be directly reflected neither in the proposed methods nor in their implementation. Addressing these scientific questions is relevant to a wide spectrum of robotic platforms and applications.

2.4 Stratification

In order to systematically design the vision model coupling system by comprehensibly introducing and arranging the underlying concepts with their interrelations, the following architecture for model-based environmental visual perception is proposed (see Fig. 2.11). The stratification is the natural embedding of spaces in a hierarchal structure. In this work, stratification is the generalization embracing spaces beyond visual reconstruction. In other words, it is the hierarchal embedding (see the upper part of Fig. 2.11) of the (a) physical, (b) visual (with perspective, affine, metric subspaces), (c) egocentric and (d) world model representation space for humanoid robots. This concept is based and extends the influential ideas presented in [63] for the visual stratification. Specifically, the stratification for model-based environmental visual perception for humanoid robots involves four heterogeneous interrelated spaces. The delimitation of each space allows a better understanding of the information flow from signals to symbols, the intermediate storage and the whole execution cycle.

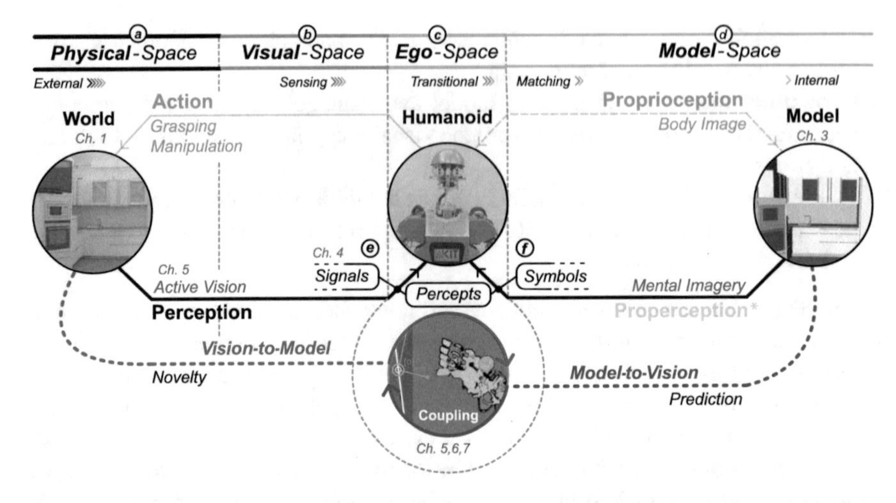

Fig. 2.11 The stratification for model-based environmental visual perception for humanoid robots. The light gray elements of the stratification are not handled in this work. Their inclusion allows a better comprehension of the nature and scope of the visual perception for humanoid robots (see [64])

- **Physical space**: This is the most concrete space. It depicts the segment of the tangible reality (stationary objects) which has been modeled and stored in the world model representation. Concretely, it includes the humanoid robot, furniture, rooms and building of the application environment.
- **Visual space**: This space embraces the projection of the physical space to the inner spatial *percepts*.[3] This is done by means of optical sensor transformations and *active vision*[4] components. The visual scene information *"streams"* inside this space forming a transducer between the physical space and this sensor-dependent space. Visual sensing, feature extraction and 2D-to-3D percept composition take place in this space.
- **Ego space**: The ego space is the short-term spatial storage for 3D percepts and global self-localization. It is a container space which stores the visual percepts obtained from different viewpoints (3D structures resulting view fusion) by controlling the robot actuators.
- **Model space**: The model space is a long-term storage containing the geometrical and topological description of the physical space entities. The indexing and hypothesis structures used for model matching and pose estimation are represented in this space.

[3]Percept is the perceptional input object, a mental impression of something perceived by the senses, viewed as the basic component in the formation of concepts (see cognition within artificial intelligence in [65]).

[4]Active vision systems interact with the environment by controlling viewpoint, exposure and other acquisition parameters instead of passively observe the scene with a fixed configuration. Usually, active sensing operates on sequences of images rather than single image, see [66].

Bidirectional Information Flows

The stratification is conceived to identify, clarify and organize the formation of the two fundamental and complementary information flows for environmental model-based visual perception (see Fig. 2.11).

- **Vision-to-model**: This flow is the *signal* $\rightarrow$ *forward* link (see lower left red flow in Fig. 2.11). It shows the information from the physical space to the ego space. It propels the information flow from sensor signals to multiview 3D percepts by means of active vision processes. Its role is to meaningfully reduce the vast amount of information obtained from the cameras and time varying kinematic configuration of the robot. This is done by extracting significant spatial structural features which can be efficiently matched with the world model representation. The robot employs this flow to connect the physical reality with the world model by transforming the photon counts from the camera sensors into structural 3D percepts.
- **Model-to-vision**: This flow is the *symbol* $\leftarrow$ *backward* link (see lower right blue flow in Fig. 2.11). It illustrates the information flow from the model space to the ego space. Its main role is to provide the necessary information to match 3D structural percepts with their corresponding shape-primitives in the world model representation. This flow is also concerned with the association and prediction of the pose and appearance of the elements of the world during task execution. The processes driving this flow manage the symbolic and geometric information necessary to realize semantic endowed perception.
- **Vision model coupling**: Based on both information flows, the *vision* $\leftrightarrow$ *model coupling* (see lower recurring flow in Fig. 2.11) arises from the synergy produced by task driven composition and coordination of the information flows. Singularly for the environmental vision, the stratification reveals a cyclical relationship between the recognition of environmental objects and the 6D-pose of the robot [67]. Due to this view-pose interdependency, in order to solve the question *"where is the robot ?"*, it is mandatory to address the question *"what is in the surroundings of the robot ?"*.

References

1. Kato, I. 1973. Development of WABOT-1. *Biomechanism 2*, 173–214. Tokyo: The University of Tokyo Press.
2. Waseda. 2012. Humanoid History -WABOT. 2012, humanoid Robotics Institute, Waseda University.
3. Sugano, S., and I. Kato. 1987. WABOT-2: Autonomous Robot with Dexterous Finger-arm–Finger-arm Coordination Control in Keyboard Performance. In *IEEE International Conference on Robotics and Automation*, vol. 4, 90–97.
4. Konno, A., K. Nagashima, R. Furukawa, K. Nishiwaki, T. Noda, M. Inaba, and H. Inoue. 1997. Development of a Humanoid Robot Saika. In *IEEE-RSJ International Conference on Intelligent Robots and Systems*, vol. 2, 805–810.

5. Yamaguchi, J., S. Inoue, D. Nishino, and A. Takanishi. 1998. Development of a Bipedal Humanoid Robot Having Antagonistic Driven Joints and Three DOF Trunk. In *IEEE-RSJ International Conference on Intelligent Robots and Systems*, vol. 1, 96–101.
6. Brooks, R., C. Breazeal, M. Marjanović, B. Scassellati, and M. Williamson. 1999. The Cog Project: Building a Humanoid Robot. In *Computation for Metaphors, Analogy, and Agents*, ed. Nehaniv C., 52–87. Lecture Notes in Computer Science. Berlin, Heidelberg: Springer.
7. Scassellati, B. 1998. A Binocular, Foveated Active Vision System. Technical report, MIT Artificial Intelligence Lab, Cambridge, MA, USA
8. Breazeal, C., A. Edsinger, P. Fitzpatrick, and B. Scassellati. 2001. Active Vision for Sociable Robots. *IEEE Transactions on Systems, Man and Cybernetics, Part A: Systems and Humans* 31 (5): 443–453.
9. Pfeifer, R., and J. Bongard. 2006. *How the Body Shapes the Way We Think: A New View of Intelligence*. New York: A Bradford Book.
10. Inaba, M., T. Igarashi, S. Kagami, and H. Inoue. 1996. A 35 DOF Humanoid that Can Coordinate Arms and Legs in Standing up, Reaching and Grasping an Object. In *IEEE-RSJ International Conference on Intelligent Robots and Systems*, vol. 1, 29–36.
11. M. Inaba and H. Inoue, Robot Vision Server, *International Symposium Industrial Robots*, pp. 195–202, 1989.
12. Jin, Y., and M. Xie. 2000. Vision Guided Homing for Humanoid Service Robot. In *International Conference on Pattern Recognition*, vol. 4, 511–514.
13. Saegusa, R., G. Metta, and G. Sandini. 2010. Own Body Perception based on Visuomotor Correlation. In *IEEE-RSJ International Conference on Intelligent Robots and Systems*, 1044–1051.
14. Bartolozzi, C., F. Rea, C. Clercq, M. Hofstatter, D. Fasnacht, G. Indiveri, and G. Metta. 2011. Embedded Neuromorphic Vision for Humanoid Robots. In *IEEE Computer Society Conference on Computer Vision and Pattern Recognition Workshops*, 129–135.
15. Vernon, D., G. Metta, and G. Sandini. 2007. The iCub Cognitive Architecture: Interactive Development in a Humanoid Robot. In *IEEE International Conference on Development and Learning*, 122–127.
16. Metta, G., L. Natale, F. Nori, and G. Sandini. 2011. The iCub Project: An Open Source Platform for Research in Embodied Cognition. In *IEEE Workshop on Advanced Robotics and its Social Impacts*, 24–26.
17. Azad, P. 2008. Visual Perception for Manipulation and Imitation in Humanoid Robots. Ph.D. dissertation, University of Karlsruhe. ISBN 978-3-642-04229-4.
18. Welke, K. 2011. Memory-Based Active Visual Search for Humanoid Robots. Ph.D. dissertation, KIT, Karlsruhe Institute of Technology, Computer Science Faculty, Institute for Anthropometrics.
19. Thompson, S., and S. Kagami. 2005. Humanoid Robot Localisation using Stereo Vision. In *IEEE-RAS International Conference on Humanoid Robots*, 19–25.
20. Lowe, D. 2004. Distinctive Image Features from Scale-Invariant Keypoints. *International Journal of Computer Vision* 60 (2): 91–110.
21. Konolige, K. Small Vision System.
22. Kagami, S., K. Nishiwaki, J.J. Kuffner, Y. Kuniyoshi, M. Inaba, and H. Inoue. 2002. Online 3D Vision, Motion Planning and Bipedal Locomotion Control Coupling System of Humanoid Robot: H7. In *IEEE-RSJ International Conference on Intelligent Robots and Systems*, vol. 3, 2557–2562.
23. Dellaert, F., D. Fox, W. Burgard, and S. Thrun. 1999. Monte Carlo Localization for Mobile Robots. In *IEEE International Conference on Robotics and Automation*, vol. 2, 1322–1328.
24. Stasse, O., A.J. Davison, R. Sellaouti, and K. Yokoi. 2006. Real-time 3D SLAM for Humanoid Robot considering Pattern Generator Information. In *IEEE-RSJ International Conference on Intelligent Robots and Systems*, 348–355.
25. Murase, H., and S. Nayar. 1993. Learning and Recognition of 3D Objects from Appearance. In *IEEE Workshop on Qualitative Vision*, 39–50.

26. Davison, A., and N. Kita. 2001. 3D Simultaneous Localisation and Map-building using Active Vision for a Robot Moving on Undulating Terrain. In *IEEE Computer Society Conference on Computer Vision and Pattern Recognition*, vol. 1, 384–391.
27. Davison, A., I. Reid, N. Molton, and O. Stasse. 2007. MonoSLAM: Real-Time Single Camera SLAM. *IEEE Transactions on Pattern Analysis and Machine Intelligence* 29 (6): 1052–1067.
28. Okada, K., M. Kojima, Y. Sagawa, T. Ichino, K. Sato, and M. Inaba. 2006. Vision based Behavior Verification System of Humanoid Robot for Daily Environment Tasks. In *IEEE-RAS International Conference on Humanoid Robots*, 7–12.
29. Okada, K., M. Kojima, S. Tokutsu, T. Maki, Y. Mori, and M. Inaba. 2007. Multi-cue 3D Object Recognition in Knowledge-based Vision-guided Humanoid Robot System. In *IEEE-RSJ International Conference on Intelligent Robots and Systems*, 3217–3222.
30. Okada, K., M. Kojima, S. Tokutsu, Y. Mori, T. Maki, and M. Inaba. 2008. Task Guided Attention Control and Visual Verification in Tea Serving by the Daily Assistive Humanoid HRP2JSK. In *IEEE-RSJ International Conference on Intelligent Robots and Systems*, 1551–1557.
31. Brooks, R., and L. Stein. 1994. Building Brains for Bodies. *Autonomous Robots* 1: 7–25.
32. Kuffner, J., K. Nishiwaki, S. Kagami, M. Inaba, and H. Inoue. 2003. Motion Planning for Humanoid Robots. In *International Symposium on Robotics Research*, 365–374.
33. Prats, M., S. Wieland, T. Asfour, A. del Pobil, and R. Dillmann. 2008. Compliant Interaction in Household Environments by the Armar-III Humanoid Robot. In *IEEE-RAS International Conference on Humanoid Robots*, 475–480.
34. Wieland, S., D. Gonzalez-Aguirre, T. Asfour, and R. Dillmann. 2009. Combining Force and Visual Feedback for Physical Interaction Tasks in Humanoid Robots. In *IEEE-RAS International Conference on Humanoid Robots*, 439–446.
35. Gonzalez-Aguirre, D., S. Wieland, T. Asfour, and R. Dillmann. 2009. On Environmental Model-Based Visual Perception for Humanoids. In *Progress in Pattern Recognition, Image Analysis, Computer Vision, and Applications*. Lecture Notes in Computer Science, eds. Bayro-Corrochano, E., and J.-O. Eklundh, vol. 5856, 901–909. Berlin, Heidelberg: Springer.
36. Harris, C., and M. Stephens. 1988. A Combined Corner and Edge Detector. In *Alvey Vision Conference*, 147–151. Manchester, UK.
37. Noskovicova, L., and R. Ravas. 2010. Subpixel Corner Detection for Camera Calibration. In *MECHATRONIKA, International Symposium*, 78–80.
38. Shi, J., and C. Tomasi. 1994. Good Features to Track. In *IEEE Computer Society Conference on Computer Vision and Pattern Recognition*, 593–600.
39. Pérez, P., C. Hue, J. Vermaak, and M. Gangnet. 2002. Color-Based Probabilistic Tracking. In *European Conference on Computer Vision-Part I*, 661–675. Berlin: Springer.
40. Canny, J. 1986. A Computational Approach to Edge Detection. *IEEE Transactions on Pattern Analysis and Machine Intelligence* 8 (6): 679–698.
41. Flint, A., C. Mei, I. Reid, and D. Murray. 2010. Growing Semantically Meaningful Models for Visual SLAM. In *IEEE Conference on Computer Vision and Pattern Recognition*, 467–474.
42. Wen, F., X. Chai, Y. Li, W. Zou, K. Yuan, and P. Chen. 2011. An Improved Visual SLAM Algorithm based on Mixed Data Association. In *World Congress on Intelligent Control and Automation*, 1081–1086.
43. Civera, J., D. Galvez-Lopez, L. Riazuelo, J.D. Tardos, and J.M.M. Montiel. 2011. Towards Semantic SLAM using a Monocular Camera. In *IEEE-RSJ International Conference on Intelligent Robots and Systems*, 1277–1284.
44. Ahn, S., M. Choi, J. Choi, and W.K. Chung. 2006. Data Association Using Visual Object Recognition for EKF-SLAM in Home Environment. In *IEEE-RSJ International Conference on Intelligent Robots and Systems*, 2588–2594.
45. Bay, H., T. Tuytelaars, and L.V. Gool. 2006. Surf: Speeded up Robust Features. In *European Conference on Computer Vision*, 404–417.
46. Jeong, W.Y., and K.M. Lee. 2006. Visual SLAM with Line and Corner Features. In *IEEE-RSJ International Conference on Intelligent Robots and Systems*, 2570–2575.
47. Klank, U., D. Pangercic, R. Rusu, and M. Beetz. 2009. Real-time CAD Model Matching for Mobile Manipulation and Grasping. In *IEEE-RAS International Conference on Humanoid Robots*, 290–296.

48. Rusu, R.B., G. Bradski, R. Thibaux, and J. Hsu. 2010. Fast 3D Recognition and Pose Using the Viewpoint Feature Histogram. In *IEEE-RSJ International Conference on Intelligent Robots and Systems*.
49. Meeussen, W., M. Wise, S. Glaser, S. Chitta, C. McGann, P. Mihelich, E. Marder-Eppstein, M. Muja, V. Eruhimov, T. Foote, J. Hsu, R. Rusu, B. Marthi, G. Bradski, K. Konolige, B. Gerkey, and E. Berger. 2010. Autonomous Door Opening and Plugging in with a Personal Robot. In *IEEE International Conference on Robotics and Automation*, 729–736.
50. Sturm, J., K. Konolige, C. Stachniss, and W. Burgard. 2010. 3D Pose Estimation, Tracking and Model Learning of Articulated Objects from Dense Depth Video using Projected Texture Stereo. In *Workshop RGB-D: Advanced Reasoning with Depth Cameras at Robotics: Science and Systems*.
51. Muja, M., R.B. Rusu, G. Bradski, and D.G. Lowe. 2011. REIN - A Fast, Robust, Scalable Recognition Infrastructure. In *IEEE International Conference on Robotics and Automation*, 2939–2946.
52. Kakiuchi, Y., R. Ueda, K. Okada, and M. Inaba. 2011. Creating Household Environment Map for Environment Manipulation using Color Range Sensors on Environment and Robot. In *IEEE International Conference on Robotics and Automation*, 305–310.
53. Chin, R., and C. Dyer. 1986. Model-based Recognition in Robot Vision. *ACM Computing Surveys* 18: 67–108.
54. Ullman, S. 2000. *High-Level Vision, Object Recognition and Visual Cognition*. The MIT press. ISBN-10: 0-262-71007-2.
55. Lepetit, V., and P. Fua. 2005. Monocular Model-Based 3D Tracking of Rigid Objects: A Survey. *Foundations and Trends in Computer Graphics and Vision* 1 (1): 1–89.
56. Roth, P.M., and M. Winter. 2008. Survey of Appearance-based Methods for Object Recognition. Institute for Computer Graphics and Vision, Graz University of Technology, Austria, Technical Reports.
57. Kragic, D., A. Miller, and P. Allen. 2001. Real-time Tracking Meets Online Grasp Planning. In *IEEE International Conference on Robotics and Automation*, vol. 3, 2460–2465.
58. Azad, P., T. Asfour, and R. Dillmann. 2006. Combining Appearance-based and Model-based Methods for Real-Time Object Recognition and 6D Localization. In *IEEE-RSJ International Conference on Intelligent Robots and Systems*, 5339–5344.
59. Azad, P., T. Asfour, and R. Dillmann. 2009. Combining Harris Interest Points and the SIFT Descriptor for Fast Scale-Invariant Object Recognition. In *IEEE-RSJ International Conference on Intelligent Robots and Systems*, 4275–4280.
60. Wei-Hsuan C., H. Chih-Hsien, T. Yi-Che, C. Shih-Hung, Y. Fun, and C. Jen-Shiun. 2009. An Efficient Object Recognition System for Humanoid Robot Vision. In *Joint Conferences on Pervasive Computing*, 209–214.
61. Azad, P., D. Muench, T. Asfour, and R. Dillmann. 2011. 6-DoF Model-based Tracking of Arbitrarily Shaped 3D Objects. In *IEEE International Conference on Robotics and Automation*.
62. Davison, A. 1998. Mobile Robot Navigation Using Active Vision. Ph.D. dissertation, Robotics Research Group, Department of Engineering Science, University of Oxford.
63. Faugeras, O. 1995. Stratification of 3-D vision: Projective, Affine, and Metric Representations. *Journal of the Optical Society of America A* 12: 46 548–4.
64. Gonzalez-Aguirre, D., T. Asfour, and R. Dillmann. 2011. Towards Stratified Model-based Environmental Visual Perception for Humanoid Robots. *Pattern Recognition Letters* 32 (16): 2254–2260. (advances in Theory and Applications of Pattern Recognition, Image Processing and Computer Vision).
65. Russell, P.N.S. 1995. *Artificial Intelligence: A Modern Approach*. Prentice Hall Series in artificial intelligence. ISBN 9780136042594.
66. Daniilidis, K., and J.-O. Eklundh. 2008. 3-D Vision and Recognition. In *Springer Handbook of Robotics*, ed. B. Siciliano, and O. Khatib, 543–562. Berlin: Springer.
67. Gonzalez-Aguirre, D., T. Asfour, E. Bayro-Corrochano, and R. Dillmann. 2008. Model-based Visual Self-localization Using Geometry and Graphs. In *International Conference on Pattern Recognition*, 1–5.

Chapter 3
World Model Representation

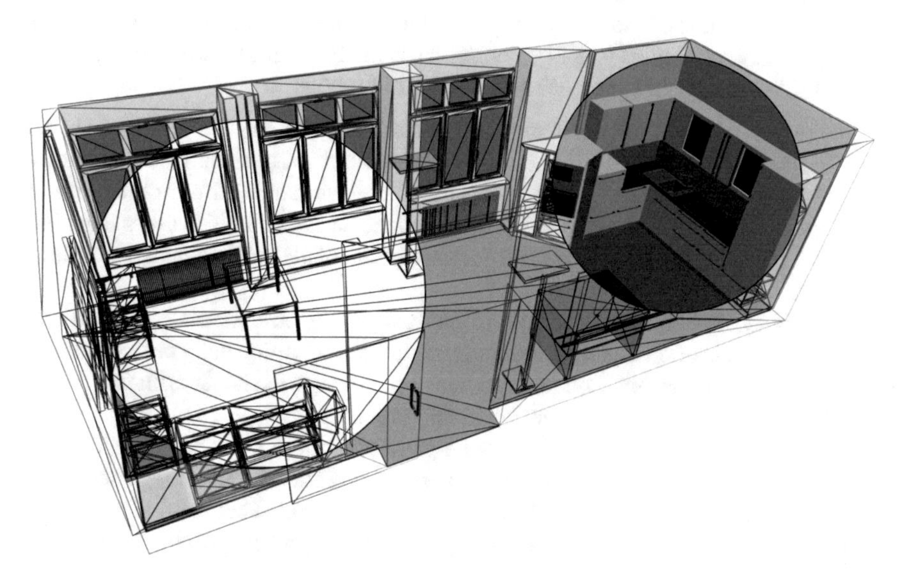

The process of visual perception for humanoid robots requires two interdependent representations. First, the *world model representation*—the one discussed in this chapter—is concerned with the formal description of the physical entities found in the application domain. Second, the *visual-feature representation* focuses on image acquisition and extraction of structural saliencies. Both representations must be highly connected in terms of similarity for association during detection and matching processes. In order to properly establish a link between these representations, the world model has to consists of concise descriptions of object *shapes*[1] arranged according to their spatial relations. Thus, the world model representation should be implemented in a hierarchical structure reflecting the spatial layout. In a complemen-

[1] *"Shape will mean the geometry of a locus of points, which will typically be the points on the surface of a physical object, or the points on an intensity edge or region in an image"*. Extracted from [1].

D. I. González Aguirre, *Visual Perception for Humanoid Robots*,
Cognitive Systems Monographs 38, https://doi.org/10.1007/978-3-319-97841-3_3

tary manner, the visual-feature representation requires a sensor-dependent language which allows the formal description of the object shapes found in the images.

In the following sections, representation criteria, viewpoint dependency and shape-primitive selection are introduced. The presentation considers the restrictions found in the visual perception process for physical interaction in terms of localization, grasping and manipulation.

3.1 Representation Criteria

An appropriate world model representation for visual recognition must satisfy various criteria in order to efficiently establish the association between visual-features and shape-primitives. These criteria allow the qualitative comparison and ranking of diverse representations (see [2]). Particularly, the following criteria have a fundamental role in the visual perception for humanoid robots:

- **Scope**: A representation has to provide means to describe a broad variety of shapes with adjustable granularity. The scope of this work lies on the human-centered environments including a wide spectrum of objects ranging from rather simple doors up to complex structures, for instance, the basket of a dishwasher machine. The scope also includes fixed and far distant elements such as walls, furniture, ceiling lamps, etc.
- **Sensitivity**: The representation has to preserve subtle differences between similar objects. This is of crucial importance for reliable recognition and pose estimation. The proposed representation must address a wide range of shapes with fine details without reducing the storage or processing efficiency.
- **Uniqueness**: The shape description of a particular object must be identified from different views for comparison. This criterion implies that another instance of the same object will produce an *"identical"* representation. This restriction enables the comparison of shapes. This assertion must be hold with respect to a comparison function.[2]
- **Stability**: This criterion requires that a slightly change in the shape of an object will produce changes with proportional magnitude in its representation. This is a desirable characteristic for recognition and a fundamental for classification (see [3]).
- **Efficiency**: This is a core performance criterion of the world model representation. There are two aspects to be considered; processing and storage. Processing efficiency depends on: (i) The computational complexity required to generate the corresponding representation (data structures) from an input shape. (ii) The complexity necessary to recall shape-primitives for matching. The storage efficiency depends on the amount of space necessary to store the generated shape-primitives. The total complexity of a representation is the result of the processing complexity

[2] A function that can filter slightly nuances produced by the inherent numerical and ordering effects of the computational representation.

plus the storage complexity. Usually, the total complexity is balanced in a trade-off to reduce the overall workload at the critical phase of recognition.

- **Scalability**: A representation capable to progressively incorporate elements in a wide range without decreasing its performance is a central key for robot applications. Therefore, the proposed mechanism for data recalling most integrate new objects without entirely recomputing all previously stored data structures. Notice that after the insertion, modification or removal of elements in the world model representation, an inherent amount of re-indexing is ineluctable. Thus, the proposed indexing schema is critical for on-board applicability and long-term autonomy (see [4]).
- **Extensibility**: Humanoid robots use different sensing modalities, for example, vision, haptics, sound, etc. Each of those modalities requires a particular information source provided by a suitable representation. This fact muss be taken into account by enabling dynamic extensions of the world model representation through forward compatibility of attributes and data structures. For instance, the world model representation could be extended from its geometric and topological structure to a cognitive representation by integrating semantic and functional attributes for symbolic planning (see [5]). The key to fulfill this criterion is to incorporate abstract and generic indexed attributes which are concretely and coherently implemented according to the particular extension of sensing modalities or recognition approaches.
- **Stochastic furnishment**: The application of probabilistic methods for robotics (see [6]) provides various important advantages. For instance, the optimal integration of different sources of information (weighted according to their certainty) enables better adaptation and improves the robustness of the systems (see [7]). This requires that the underlying representation provides means to manage dynamic probabilistic attributes. Therefore, a comprehensive representation must include such stochastic mechanisms.

3.2 Viewpoint Dependency

The spatial relation of the elements composing an object requires a reference frame. Its selection determines the data structures to be stored. This selection also circumscribes the searching methods. In the literature (see surveys [1, 8]), there is a pervasive twofold division of reference frames:

- **Object-centered**: The selection of a single global frame attached to the object constitutes the so-called primitive-based representations.
- **Viewer-centered**: These are appearance-based representations with multiple local frames associated with viewpoints form which the object is captured (usually by images) in order to be represented.

Both representation categories have advantages and disadvantages which usually act complementary to each other (see [9]). For instance, object-centered representations are concise and accurate in terms of metrics while viewer-centered representations depend on the resolution of the captured or rendered images. Another complementary property is redundancy. Viewer-centered representations contain a high degree of redundancy while the object-centered descriptions are condensed.

The view dependency of the application domain narrows the selection of the representation. In the case of environmental visual perception, the following aspects must be considered:

- **Pose**: A view-based representation is a partial summary of the appearance of the object. Therefore, stereotypic viewpoints have higher representativeness and provide better visual-features for matching (see [10]). This implies critical biased spatial distributions of the recognition results. This condition is imposed by the relative 6D-pose between the camera(s) and the object(s). The distribution of the recognition rate is negatively biased because there are configurations where the appearance-based recognition methods systematically fail despite of the sufficiently (conclusive geometric) information necessary to recognize and assert both the identity and pose of the object. This is a key argument to avoid viewer-centered representations for environmental visual perception. On the contrary, based on object-centered representations, it is plausible to recognize and estimate the pose of the object (even from non-stereotypical views) as long as there is enough geometric information available in the scene.
- **Composition**: Both representation categories are limited in terms of object composition. Object-centered representations can hardly store all possible attributes with sufficient granularity and precision without incurring in storage efficiency issues. Viewer-centered representations also present these and other even more restrictive processing issues. A remarkable example depicting these problems is the viewer-centered representation called *aspect graphs* [11]. They are collections of distinctive image nodes which are connected by links to other image nodes sharing common spatial region(s) of the object. Determining the minimal set of image nodes which soundly capture the structure of the object established a research field. This research field was abandoned after the contribution of S. Petitjean et al. in [12]. In that work, it was shown that under perspective transformation the complexity of the aspect graph of (rather simple) smooth algebraic surfaces is of order n^{18}, where n stands for the surface degree.[3] When using viewer-centered representations, the amount of views grows at an intractable rate. This is a reason to avoid viewer-centered representations. In contrast to such enormous complexity, object-centered representations have superior properties in terms of object *aggregation*.[4] The independence of the viewpoint enables the exploitation of object-centered representations in scenarios where objects may partially occlude each other. For instance, when a humanoid robot obtains the representation of a cupboard, the particular structure and appearance of the door is not associated with the representation of the handle. Thus, the feature extraction system and the world representation are

[3]The algebraic surface order n is the degree of the polynomial shape of the surface, namely the maximum intersection points between a straight line and the surface.

[4]The physical attachment of objects by rotational or prismatic joints.

capable of properly working together in the presence of diverse materials, colors or geometric compositions of the objects. This is key argument for selecting an object-centered representation for visual perception in humanoid robots.

- **Occlusion**: The partial visibility of the objects is produced by three facts: (i) Self-occlusion occurs in all opaque objects in the world. (ii) The field of view is not only limited by the photometric precision of the optical system but also by the base-line length in the head of the humanoid robot. (iii) In application where various objects have to be detected, the occlusion produced by other objects is omnipresent and unavoidable. The partial visibility produced by self-occlusion is implicitly considered within viewer-centered representations. However, during recognition there are four strong assumptions to be fulfilled for these descriptions to work: (i) Objects can be conclusively segmented from the background. (ii) Objects experience only self-occlusion. (iii) Objects are entirely contained in the image. (iv) It is possible to conduct a vast sensor and instance-dependent learning stage in advance. Since these restrictions clearly exclude the conditions in environmental visual perception for humanoid robots, it is necessary to reflect upon the effects of partial visibility within object-centered representations. Due to their nature, an object-centered representations cannot be easily endowed with an occlusion-tolerant mechanism. There are approaches [13] concerned with the visibility analysis in object-centered representations. However, this is still an active research field which can be reasonably mitigated by local means. The practical solution to the occlusion problem lies in the use of local shape features which—despite being partially occluded—still allow robust object recognition (see Sect. 3.3).
- **Transformation**: The physical interaction with the environment in terms of grasping and manipulation by humanoid robots implies rigid and non-rigid[5] transformations reflected in changes of pose and structural constitution of objects. Due to the nature of the object-centered representations, it is possible to describe hierarchical organizations of objects by parametric 6D rigid transformations. In order to realize such dynamic spatial arrangements, a *scene graph*[6] is commonly used (see an interesting example which visualizes the uncertainty in the transformations in [16]). The scene graph contains efficient multiple resolution data structures, space partitioning and traversal methods necessary to describe environments, enable photorealistic visualizations [17] and enable physical simulations [18]. These approaches have been widely studied and extensively used in medical domains, robotics (see [19, 20]), scientific simulators, video games and augmented reality applications just to name a few. In contrast, viewer-centered representations cannot expeditiously cope with the hierarchical organization of the objects under parametric transformations. Moreover, viewer-centered are not endowed with extensibility means for various important aspects such as physical simulations, efficient on-demand generation of

[5]Non-rigid transformations are beyond the scope of this work (see approaches on representation and recognition of non-rigid objects and transformations [14, 15]).

[6]A data structure used to hierarchically organize and manage the contents of a spatially organized scene (see example in http://www.coin3d.org).

novel images from different viewpoints or semantic enrichment for planning and reasoning.

- **Accuracy**: Object-centered representations are explicit, attribute-based, symbolic and metric descriptions. Therefore, their accuracy and precision depend on their numerical resolution (for example, using floating point real numbers [21]) and the assumption of a closed-world [22]. Consequently, assertions in these representations are accurately done because the visual measurement resolution is lower than the numerical representation. In contrast, viewer-centered representations are based on exemplary manifestations of the objects. Hence, the information only implicitly stored is not accurate outside these exemplary viewpoints. Particularly in image-based representations, the perspective transformation, lighting effects and dynamic range of the devices (physical or simulated) used to obtain the images notably restrict the visual measurement resolution. Accordingly, a visual assertion based on a viewer-centered representation hardly can be more accurate than the minimal spatial resolution of a pixel. This is another key argument for selecting object- instead of a viewer-centered representations in the context of vision-based localization, grasping and manipulation.
- **Matching**: An additional consequence of the reference frame is the establishment of a proper association space necessary to match visual-features against shape-primitives. These spaces can be categorically separated according to [2] into *correspondence space*, *transformation space* and *hybrid spaces* as combinations of both. In the correspondence space, the matching of visual-features is done by relational expansion of association nodes. These nodes are representations linking visual-features to shape-primitives. For example, the *interpretation tree* [23] is an important foundation of the matching methods based on the correspondence space. However, due to the combinational complexity (even with accelerated extensions [24]), its straightforward application is prohibitive in cluttered scenes. In the transformation spaces, visual-features are mapped into a tailored *parametric voting space* or a *prototypical characteristic space*. In these spaces, the visual-features are matched with shape features in very efficient manners. In voting spaces, matching is realized by selecting the highest voting bin or cluster from which coordinates it is possible to attain the corresponding match. Widespread examples of these transformation spaces are the *generalized Hough transform* [25], *geometric hashing* [26] and *tensor voting* [27]. In the characteristic spaces, matching is performed by classifiers such as nearest-neighbor [28], support vector machines [3] or various neural network approaches [29]. Due to its nature, the matching process operating on viewer-centered representations is restricted to classification, clustering or projection to special tailored spaces [30]. In contrast, object-centered representations are not limited to a particular matching schema. This occurs because explicit representations allow the matching to be conducted in all these spaces, namely at the attribute level in corresponding space or, alternatively, the matching process can be done in an expedient prototypical space designed for a particular purpose (see [31]). In these cases, the object-centered representation is mapped into a prototypical space in order to realize matching by classification. In summary, the flexibility

of object-centered representation is very convenient because it allows the integration of heterogeneous matching methods with diverse sensing modalities in a wide spectrum robotic applications.

3.3 Shape-Primitives: Related Work

Based on previously discussed viewpoint dependency and representation criteria, the world model addressed in this work is an object-centered representation. This frame of reference partially delineates the structure of the world model representation. However, the complete outline must also consider other aspects which depend on the particular shape descriptor of the shape elements of the object representation, namely, the *shape-primitives*. These atomic elements are usually selected according to the particular sensing modality and matching approach. In order to state founded arguments about the particular selection of shape-primitives in this work, a basic survey on shape-primitives for visual recognition is presented.

3.3.1 Blocks-World

In visual perception, light and relative poses are cofounded in the appearance of the objects. The extraction of visual-features from this cofounded manifestations captured by images can be robustly characterized by the discontinuities (in structure, chrominance or luminance) of the surface of the objects. These discontinuities are usually called (geometric, color or intensity) edges. They provide a high degree of invariance of viewpoint and lighting effects compared to image patches or other visual-features (see [2]). This happens in a remarkable manner while analyzing single color polyhedral with homogeneous color with high-contrast background illuminated by controlled lights. In such setups, the geometric edges of an object are so close related to its visual edges that the necessary mapping between the object-centered representation and the visual-feature representation is trivial (see Fig. 3.1). This simplification was called *"blocks-world"* [32]. These setups were used during the initial stages of artificial vision. The research on blocks-world ended with the *MIT demo copy* [33]. In this demonstration, a robot was capable to reproduce a given scene by visual analysis and manipulation using a hand-eye calibrated system. There were various seminal contributions provided by the *MIT Robot* including visual-feature extraction, pose estimation and grasping methods. Despite the insights gained through these approaches, the blocks-world scenarios were far from the solution of free form objects, occlusions, shadows, textures, etc.

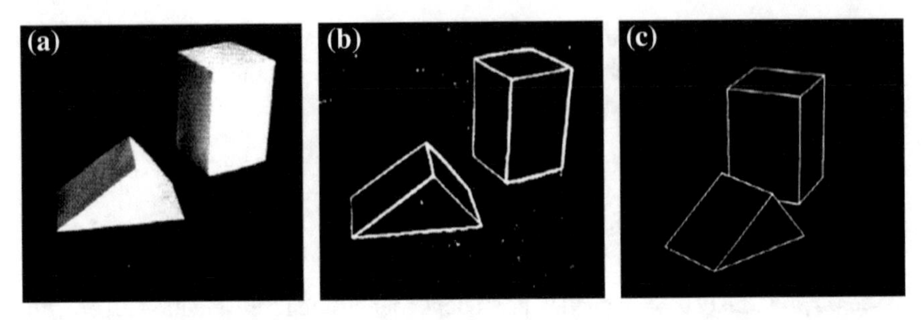

Fig. 3.1 The *"blocks-world"* setup, extracted from [32]. **a** Source scene with two polyhedral objects. **b** Extracted edges by the Roberts gradient operator. **c** A novel viewpoint of the scene. [©1963 - MIT Press]

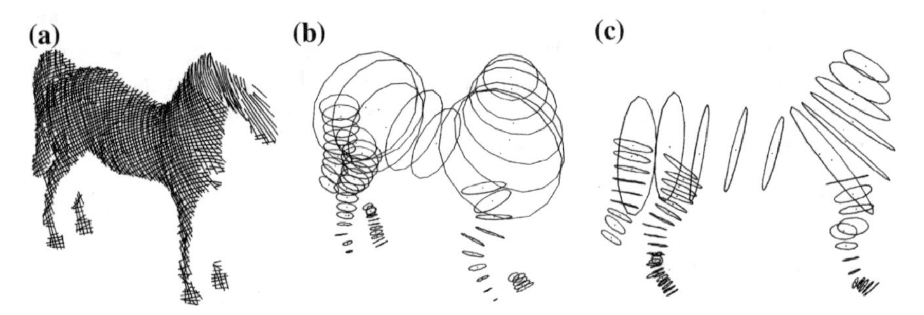

Fig. 3.2 Generalized-cylinders. **a** Contour lines of a toy horse acquired by laser projection and captured by television cameras [35]. **b** Results of the range data analysis compactly represented by circles along the transformation axis. **c** A different viewpoint of the scene. [©1973 - IEEE]

3.3.2 Generalized Cylinders

The search for object representations for free-form objects brought diverse extensions of the blocks-world. The first extensions were curved edges. The contribution of the work of A. Guzmán et al. in [34] was the introduction of spatial constrained parts. However, the nature of the line drawings used in this method made it unfeasible for complex applications. Subsequently, a more flexible representation based on space curves and circular cross-section functions called *generalized cylinders* was proposed by G. Agin et al. [35] (see Fig. 3.2). In this representation, the objects are described as segments produced by grouping parallel point traces obtained from laser range systems. Further research in this direction allowed M. Zerroug et al. [36] to apply generalized cylinders using intensity images. Despite the success in these simple and other complex military scenarios, the generalized cylinders lacked of wide representation scopes and presented high instability. Additionally, generalized cylinders were not appropriate to manage occlusion nor complex lighting setups found in everyday applications. Nevertheless, the formal inclusion of analytical constraints within the shape-primitive representation appeared for the first time in the generalized cylinders. This key concept enabled the matching of visual-features to

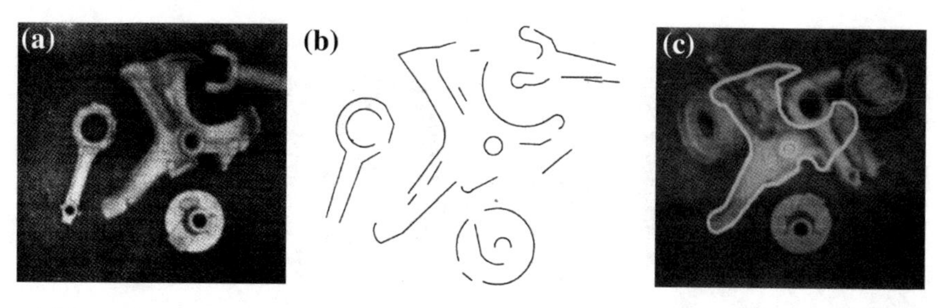

Fig. 3.3 The model-based representation of W. Perkins for visual object recognition [37]. **a** Example of a digitized input image. **b** Intermediate results of the image processing pipeline illustrating lines and edge points. **c** The model superimposed on the gray-scale image shows the robustness of the system even with the presence of occlusions and complex clutter. [©1978 - IEEE]

shape-primitives with the help of a solid theoretical body of computational geometry. Since the circles along the axis had to be obtained by fitting noisy and occluded data. The low stability and high sensitivity of the representation were limited. Thus, this restricted the generalized cylinders application.

3.3.3 Model-Based Representations

With a different motivation, geometric constraints have also been proposed into model representations in industrial applications (at the General Motors) within the work of W. Perkins et al. in [37] (see Fig. 3.3). With this approach, a machine vision system was developed to estimate the pose of complex curved objects using intensity images and geometric models. The so-called *concurves* models were generated and analyzed offline in order to find the visual-features which were likely to be found in the images.

This key concept had significant consequences for the following representations including this work. It made clear that pure data driven (bottom-up) approaches are less capable than those which consider each model separately and strategically use its representation to plan an optimized search of visual-features in a top-down approach. Although this model-based representation overcomes many of the limitations of previous representations, it had still intrinsic limitations as well. These limitations were: (i) Strong assumptions of the relative pose between camera and objects. (ii) The representation restricted objects having a dominant plane with salient boundary edges, namely the concurve features.

3.3.4 View Constraints and Perceptual Organization

Later on, the model-based representations were enhanced by considering the image formation process. Specifically, the effects of projection from 3D scenes to 2D

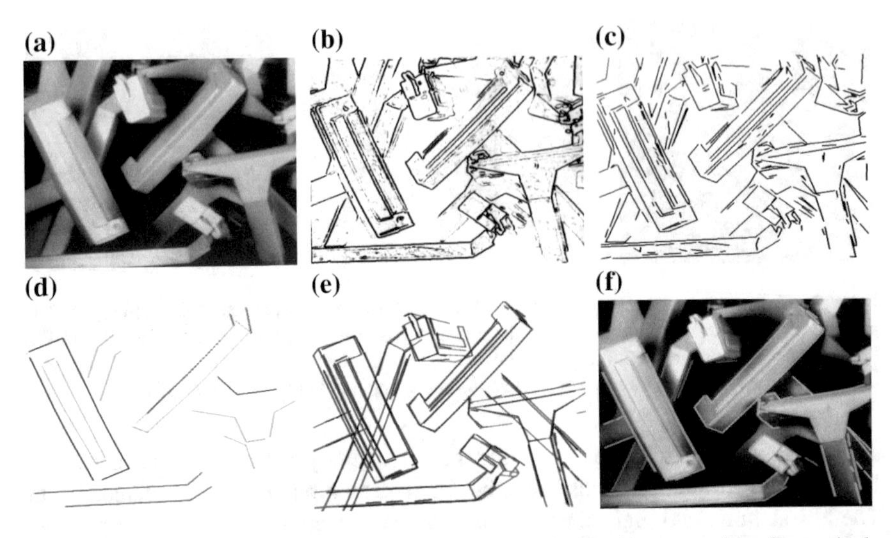

Fig. 3.4 The *SCERPO* system, a model-based vision system for 3D object recognition from a single 2D image by D. Lowe [38]. **a** Input image of a bin with disposable razors. **b** Edge pixels resulting from Laplacian convolution. **c** Straight lines from a scale-independent segmentation algorithm. **d** Highly ranked perceptual groups of line segments using Gestalt concepts such as parallelism, collinearity and proximity of endpoints. **e** Superimposition of matches between line segments and consistent viewpoints. **f** Recognition results are displayed (considering occlusion) on top of the input image. [©1987 - Elsevier]

images. The first outstanding work to present a systematic approach of projective consistency constraints using models was introduced in the *SCERPO* system [38] (see Fig. 3.4). Similar to the work of W. Perkins, the recognition and pose estimation of the approach of D. Lowe [38] has been done in a top-down manner. It was performed by the application of the perceptual organization founded on *Gestalt*[7] principles:

"Perceptual organization refers to the ability to impose organization on sensory data, so as to group sensory primitives arising from a common underlying cause. The existence of this sort of organization in human perception, including vision, was emphasized by Gestalt psychologists. Since then, perceptual organization ideas have been found to be extremely useful in computer vision, impacting object recognition at a fundamental level by significantly reducing the complexity." Extracted from [40].

The key idea of the SCERPO system was to use various Gestalt perceptual criteria (parallelism, collinearity and proximity of endpoints) to group line segments in constrained configurations. These constrained configurations correspond to object models which are likely to be invariant over a wide range of viewpoints. In order to reduce the combinational explosion during the feature matching, a novel probabilistic ranking and grouping was introduced. The influence of this work can be

[7] See a survey of up-to-date methods on perceptual organization in [39].

seen in most of today visual recognition approaches. In particular, the two important contributions are:

- **Gestalt geometric patterns**: The reinforcement of structural regularity by means of Gestalt principles is a suitable strategy in man-made object recognition.
- **Single viewpoint projection**: The coupling constraints between visual-features and shape-primitives using viewpoint consistency are fundamental for robust recognition and efficient validation.

Despite the advances provided by the viewpoint constraints and perceptual organization, the SCERPO system had limitations as well. The geometric primitives used in the representation were limited to line segments. The amount and quality of the line segments were strongly limited by the edge extractors available at that time. Due to the interest in having general geometric-primitives and perceptual grouping schemes, this research split in two directions: (i) Recognition in terms of projective constraints denominated *"alignment problem"*. (ii) Recognition by intrinsic shape properties called *"geometric invariants"*.

3.3.5 Visual Alignment

In case of visual recognition is considered as an alignment problem, the visual-features are fitted[8] to the shape-primitives of the object representation. This definition and approaches can reduce the complexity of shape-primitives to points, point pairs or line segments. For example, the work of D. Huttenlocher et al. [41]. In that approach, the transformation from 3D model points to 2D image points is done using edges connected by point pairs found along the contour of the object (see Fig. 3.5).

In order to prevent mismatches, the complete contour of the object was tracked and the estimated pose was verified for consistency. These concepts have been implemented into an affine voting schema which allows object recognition in cluttered and partially occluded scenes. Despite of the high robustness achieved (about 99% success recognition according to [43]), their assumption of an affine transformation (weak perspective instead of full projective transformation) and the high computational complexity (in textured images or complex models) restricted its wide application.

3.3.6 Geometric Invariants

An alternative visual recognition approach is to use computationally inexpensive properties which capture the shape of the object independently of its pose. This allows pose invariant comparison of objects. These properties are called *geometric*

[8] A rigid transformation for minimizing an objective alignment function is estimated.

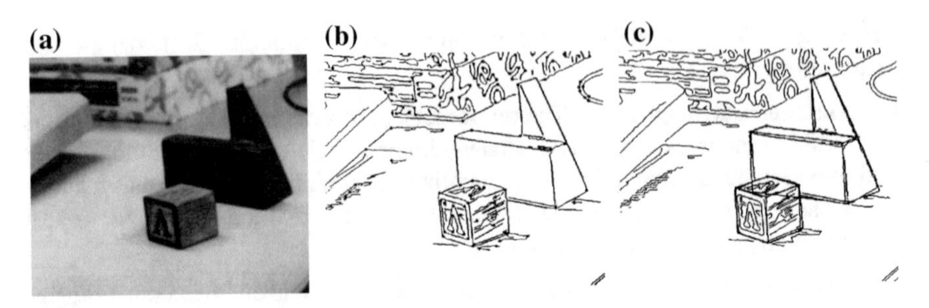

Fig. 3.5 Edges connected by point pairs, a representations for visual alignment [41]. **a** Test input image for visual object recognition. **b** Detection of edge segments using *Canny* filter [42]. **c** Result of the recognized instances shown by overlayed lines. [©1990 - Springer]

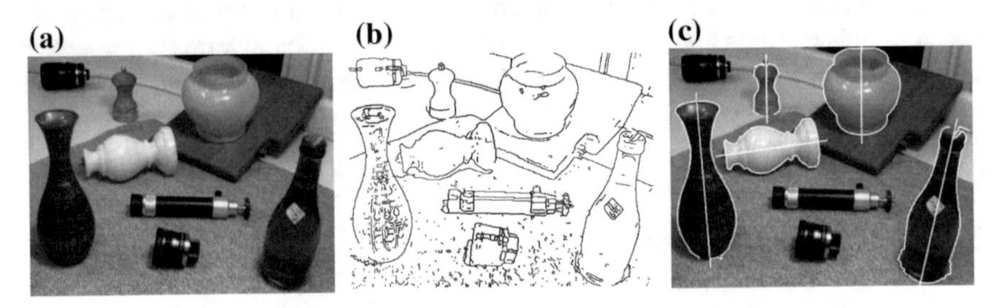

Fig. 3.6 Geometric invariants in solids of revolution [46]. **a** Solid objects with diverse revolution profiles. **b** Extracted linked edge points. **c** The resulting revolution surfaces with superimposed axes. [©1990 - Oxford University Press]

invariants. In these approaches, the recognition consists of finding projective invariants in the images which are stored as characteristic model properties. These ideas have been deeply analyzed. The most important contribution has been the discovery: *"there is no viewpoint invariant for general 3D shapes"*, (see [44, 45]). Following this lack of generalization, the research focused on a visual dictionary (case-based invariants) which could be used to recognize a broad scope of shapes. These ideas were successfully implemented in various systems. For instance, the work of A. Rothwell in [46] (see Fig. 3.6).

Geometric invariant approaches use small collections of lines and points as representation primitives. The main limitations of these approaches is the need for grouping during feature extraction (as in Sect. 3.3.4). Additionally, sensor noise has strong effects in the invariance matching. This occurs due to the lack of redundancy in the estimation of similarity between model invariant properties and structures found in the images. The absence of general matching schemes and the limited amount of geometric shapes with invariant features limited the research and wide application of geometric invariants. When these limitations became more apparent, a new rapidly growing tendency in the field of visual representations and matching appeared, the appearance-based methods. These methods exploit an alternative concept, the visual manifold learning (see highly influential work in [29]).

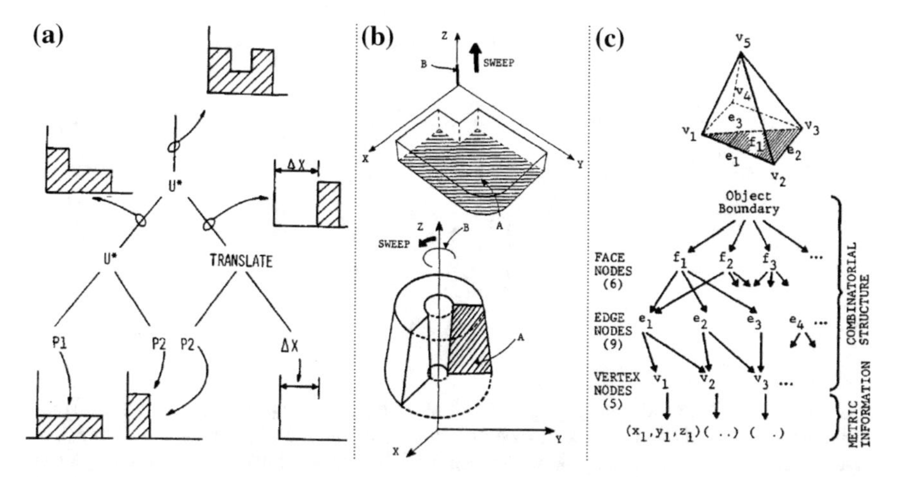

Fig. 3.7 Geometric modeling representations for solids (extracted from [47]). **a** constructive trees are formal representations of 3D models based on boolean operators (inner nodes) acting upon shape-primitives (leaves nodes). **b** two examples of translational and rotational sweep representations. **c** the boundary representation consists of the metric (vertex coordinates) and the combinatorial levels (vertex incidences) defining edges and faces as shape-primitives of the objects. [©1980 - ACM]

3.3.7 Geometric Modeling Systems: CAD-CAM

The representation primitives used in previous model-based approaches are subsets of lines, line segments or points from an over simplified geometric representation of the objects. This kind of shape-primitives dominated the research of computer vision during the first three decades. During this period, intensive research on representations was taking place in parallel within the fields of computational geometry, computer graphics and manufacturing. Computer aided design CAD and computer-aided manufacturing CAM methods have been founded in these representations [47] (see Fig. 3.7). In the work of A. Requicha et al. in [47], the mathematical properties[9] of each representation including spatial enumerations (*voxels*) and geometric-primitives, such as Platonic solids, spheres, cones, etc. were presented. Based on CAD-CAM representations, most model-based approaches use the boundary representation for visual perception. This occurs because these representations provide direct access to the shape-primitives (points, edges and surface) while keeping the connected structure of the object. This linked structure is necessary to cope with partial occlusions. The proper application of boundary representations implies, these explicitly object-centered representations are comprehensive containers in terms of geometric invariants, perceptual organization and alignment methods. In other words, the CAD-CAM representations subsume and surpass previous alignment (in Sect. 3.3.5) and invariant representations (in Sect. 3.3.6). Furthermore, the primitive incidence in boundary descriptions allows graph conceptualizations of the models.

[9]See mathematical validity, completeness, uniqueness and conciseness in [47].

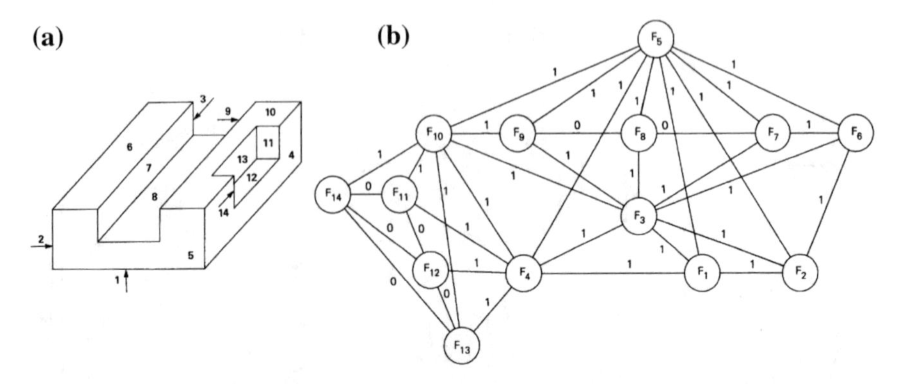

Fig. 3.8 The attributed adjacency graph (AAG) of a polyhedral object [48]. **a** The visualization of the model shows the indices used to identify the object faces. Notice the arrows for occluded faces. **b** The AAG representation shows the face incidence. This attributed adjacency graph has a unique node for every face. For every line there is a unique edge connecting two nodes sharing the common line. Every edge has a numeric attribute with value 0 if the faces sharing the edge form a concave angle or a value 1 if the faces form a convex angle. This particular use of attributes and recognition has been designed for manufacture planning. [©1988 - Elsevier]

This enables a huge body of discrete mathematics for matching visual-features to shape-primitives. These advantages are the main arguments why most recognition approaches based on CAGM (Computer-Aided Geometric Modeling) use of boundary representations.

3.3.8 Graph-Based Geometric Representations

An early work on recognition of machined surfaces from 3D solid models using boundary descriptions and graph theory was presented by S. Joshi et al. [48] (see Fig. 3.8). In that work, the authors presented the attributed adjacency graph (*AAG*) for the recognition of machined surface features from 3D boundary representations. Their contribution was a formal graph schema for surface identification based on convexity attributes. The proposed boundary representation (a bidirected graph) allowed the extraction and matching of features such as holes, slots, steps, etc. This method evolved in specific tool applications (see [49]).

3.3.9 Hybrid Representations

At the end of the geometric invariant era (see Sect. 3.3.6), the appearance-based methods dominated the research fields for object representation and matching for visual recognition. The rationale was, the feature extraction such as edge detection

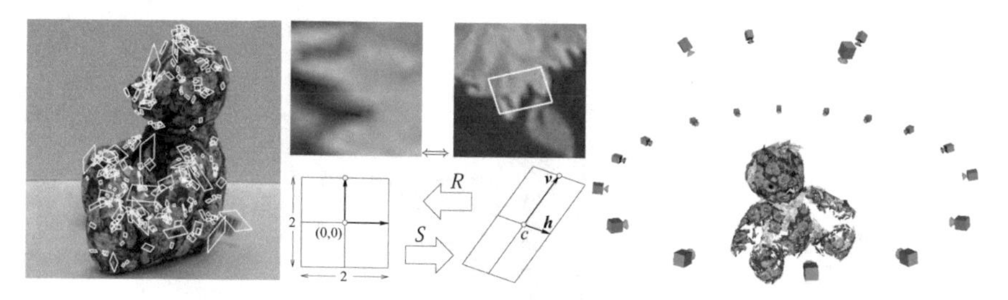

Fig. 3.9 Hybrid representation including image patches and spatial relationships. The rectified patch and its original region are illustrated by means of forward R and backward S matrices. Extracted from [52]. [©2006 - Springer]

and region segmentation were not stable enough for general object and scene recognition due to photometric effects, complex textures and non optimal lighting setups. Therefore, the community moved away from geometric primitives and started to consider image regions (small image patches) as the key for representation and matching. There were various important contributions within this research field including a large number of algorithms focusing on point-based scale invariant features and descriptors such as SIFT [50] and SURF [51]. A remarkable contribution (in terms of hybrid representations using point-based features and geometric constraints) is the work of F. Rothganger et al. [52] (see Fig. 3.9). In that work, the authors present local affine-invariant descriptors from image patches with spatial relationships. This representation provides advantages in terms of feature detection (because of the broad variety of high quality feature extractors) and structural stability (due to the geometric structure in the spatial relations of the features).

The combination of geometric information with appearance-based information allows robust and efficient recognition. This is illustrated in [53]. In this work, the authors present a markerless system for model-based object recognition and tracking (see Fig. 3.10).

This representation combines viewer-centered and object-centered characteristics. In the offline generation of the representation, various views of the object (so-called key-frames) were captured with calibrated cameras. Manual labeling was done in order to register the views. The result is a set of 3D points associated with affine image-patch descriptors. Based on this information, a 3D model of the object can be generated and used for tracking by meas of image rendering from novel viewpoints. The pose estimation used to initialize tracking has been done by the POSIT (Pose from Orthography and Scaling with ITerations) method from [54]. Finally, the tracking was robustly achieved by integration of a M-estimator (based on [55] and RANSAC [56]) along with key-frames from previous locations using an adequate adjustment method. The combined representation improves the system performance in terms of robustness and efficiency for online recognition and pose estimation compared to their previous work in [57]. The critical disadvantage of these representations is that the acquisition of labeled features requires a huge overhead for real scale applications.

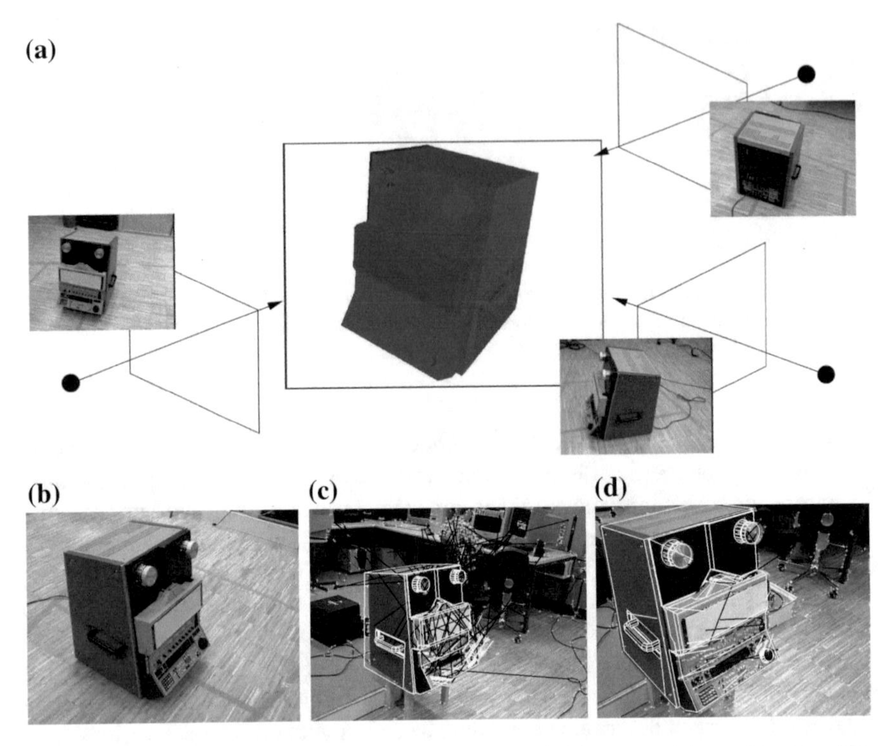

Fig. 3.10 Combined representation for model-based visual object recognition and tracking [53]. **a** The off-line capturing and labeling of key-frames is used to generate the representation including appearance and geometric structure in the form of a CAD boundary model. **b** The object to be recognized. **c** Result of the recognition with pose estimation. The multiple point-features and black lines show spurious associations. **d** A different viewpoint shows the robustness of the system. [©2003 - IEEE]

3.3.10 Qualitative Analysis

The discussed representations show a perspective of visual recognition covering areas of interest in environmental visual perception. In the literature, there are numerous model representations—mostly combinations—used for visual perception. The properties, criteria and aspects introduced in this section serve two purposes: (i) Define a catalog of concepts used along this work. (ii) Categorically compare different shape-primitives to identify the most convenient ones.

Because of the difficulties for measure all properties under compatible conditions, the evaluation in Tables 3.1 and 3.2 addresses qualitative observations. Based on this analysis, the world model representation for this work is proposed (in Sect. 3.4). Concretely, a computer aided geometric model (CAGM) based on boundary representation contained in a multilevel graph schema is used. This representation provides all the advantages of the geometric model-based representations while incorporating features for probabilistic indexing.

Table 3.1 Qualitative analysis of the representations based on shape-primitives according to the viewpoint dependency (Sect. 3.2). **Notation**: **H**igh, **M**edium, **L**ow, **U**ndefined, **N**one, **O**bject-centered, **V**iewer-centered, **R**estricted static, **S**tatic, restricted **D**ynamic, **C**orrespondence space, VoTing space, **P**rototypical space, **A** self-occlusion, **K**nown object(s) occlusion and **X** unknown object(s) occlusion. Gray cells indicate the best quality per column

Representation	Viewpoint dependency	Representable complexity	Model aggregation	Transformation capabilities	Accuracy	Matching space	Occlusion tolerance
Section 3.3.1 blocks-world	O	L	L	R	H	C	AK
Section 3.3.2 generalized cylinders	O	U	N	R	L	CP	A
Section 3.3.3 concurves	O	L	H	R	M	CP	AX
Section 3.3.4 gestalt organization	O	M	H	S	M	C	AX
Section 3.3.5 primitive alignment	O	L	H	S	M	V	AX
Section 3.3.6 geometric invariants	O	L	M	S	M	CT	AX
Section 3.3.7 CAGM: CAD-CAM graph/geometric based	VO	M	H	D	H	CP	AX
Section 3.3.9 hybrid: appearance and geometric	VO	H	N	S	M	CP	AX

Table 3.2 Qualitative analysis of the representations organized by shape-primitives (Sect. 3.3) according to the representative criteria from Sect. 3.1. **Notation**: **H**igh, **M**edium, **L**ow, **U**ndefined, **N**one. (1) None of the control-points representations such as *B-splines* or *NURBS* (see [61]) were discussed. (2) The Sect. 3.4 presents mechanisms which support the proposed representation with probabilistic features. (3) There are various approaches enabling this kind of representations with probabilistic features (see [62]). The gray cells indicate the best quality per column

Representation	Shape primitive(s)	Scope representable models	Off-line training	Sensitivity	Uniqueness	Stability	Processing	Storing	Scalability	Extensibility	Probabilistic
Section 3.3.1 blocks-world	Line segments	Polyhedral	N	L	U	H	H	H	U	U	N
Section 3.3.2 generalized cylinders	Axes and radial cross- section functions	Restricted	H	L	M	L	U	H	U	U	N
Section 3.3.3 concurves	Center points and contours	Dominant plane	N	M	L	M	H	H	U	L	N
Section 3.3.4 gestalt organization	Line segments	Faceted	N	M	M	M	M	M	L	L	H
Section 3.3.5 primitive alignment	Subset primitives: points, lines, segments	Faceted	N	M	L	L	L	M	L	U	N
Section 3.3.6 geometric invariants	Subset primitives: points, lines, segments	Faceted, R-surfaces	N	M	L	L	M	H	L	U	N
Section 3.3.7 CAGM: CAD-CAM graph/geometric based	Boundary primitives: vertices, normals, edges and faces	Subdivision surfaces, control-points[1]	N	H	H	H	H	H	H	H	H[2]
Section 3.3.9 hybrid: appearance and geometric	Combined primitives: point descriptors with spatial relationships	General	H	H	H	M	M	L	L	H	H[3]

Furthermore, the envisaged visual perception for humanoid robots requires a representation which provides explicit metric information in a compact and efficient form. Since the pioneer work on robot vision of [32], it has been shown that a geometric model for the representation of entities in the world can lead to a successful perception-planning-action cycle. More recently, this has been generalized using CAD for visual tracking while monitoring task execution [28, 53, 58]. By employing geometric models in form of CAD representations, it is possible to enable robots to recognize and localize previously unseen model instances without sensor-dependent representations.

This advantage arises from the lack of training required in appearance-based visual recognition approaches (from the pioneers in this field [29] up to the recent advances [59, 60]). Implicitly, the sensor-independence of the representation expands the potential of robots by reducing the amount of information for representation allowing the access of compact models from various sources via Internet.

Moreover, most of todays industrial *consumer packaged goods* (such as groceries), *durable goods* (instruments and tools), *major appliances* (such as furniture and electronic appliances), vehicles and even buildings and cities are being designed with CAD systems. Thus, creating models for robotic applications will be significantly supported by web technologies in the near future. A remarkable example of evidence is the public *Google 3D Warehouse*.[10] The model acceptance criteria for the so-called *"3D-Dictionary"* are very rigorous.

This results in high quality models—even with GPS registration—which can support various metric, virtual and augmented reality applications, ranging from 3D entertainment and virtual navigation to disaster management, urban logistic planning, autonomous vehicles and humanoid robot applications. Other evidence supporting the plausibility of model-based approaches are advances in automatic acquisition of environmental models with metric and functional enrichment [63–65]. These methods are based on active-sensors (non-human like modalities) and require heavy post-processing stages.

3.4 Multiple Level Graph Representation

The proposed world model representation (denoted as $\mathbb{W}$) is a multiple level graph based on novel and existing concepts. The graph representation $\mathbb{W}$ is a spatially registered and hierarchical organized collection of object models

$$\mathbb{W} := \left\{ \{O_i\}_{i=1}^{n} , \mathcal{U} := I \in \mathbb{R}^{4x4} \right\}, \tag{3.1}$$

[10]The *Google 3D Warehouse* is a free online repository designed to store, share and collaborate on 3D models, visit http://sketchup.google.com/3dwarehouse/.

where the object models O_i are also embedded graphs for them selves with geometric and topologic information described in the world model coordinate system (denoted as $\mathcal{U} \in SE^3$ expressed in homogeneous coordinates). In order to provide a sequential and deductive presentation of the multiple level graph representation $\mathbb{W}$, a spatial hierarchy is introduced (in Sect. 3.4.1) which leads to a detailed formalization of object models (in Sect. 3.4.4). Indexing mechanisms necessary to match visual-features with shape-primitives are introduced (in Sect. 3.4.5). Subsequently, important aspects involved in the data conversion from CAD interchange files to the formal representation $\mathbb{W}$ are discussed and illustrated with examples (in Sect. 3.4.6). Afterwards, the evaluation and analysis of some experimental scenarios is presented (in Sect. 3.5). Finally, a discussion about the qualities, advantages and limitations of the proposed world model representation are presented (in Sect. 3.6).

3.4.1 *Spatial Hierarchy*

The envisaged spatial hierarchy is established by the arrangement of all[11] $n = |\mathbb{W}|$ object models within the world model representation according to the *spatial enclosing principle*. The spatial enclosing of an object model O_i is defined by the subspace extraction function[12]

$$\Theta(O_i) : O_i \mapsto \mathbb{S}_i \subset \mathbb{R}^3, \tag{3.2}$$

which determines the 3-dimensional subspace $\mathbb{S}_i$ occupied by the object model. Consequently, an object model O_j is denoted to be fully contained $O_j \succ O_i$ within the object model O_i if and only if the subspace occupied by $\mathbb{S}_j = \Theta(O_j)$ is a proper[13] subspace of $\mathbb{S}_i = \Theta(O_i)$, namely

$$O_j \succ O_i \Leftrightarrow \mathbb{S}_j \subsetneq \mathbb{S}_i \Leftrightarrow \Theta(O_j) \subsetneq \Theta(O_i)\,. \tag{3.3}$$

3.4.2 *Directed Hierarchy Tree*

As consequence of the directed ranking sustained (by Eq. 3.3), the object model O_j is placed in a lower hierarchical level compared to the hierarchy level of O_i. The generalization of this containment produces an ordered relation formalized by an acyclic directed graph, namely a directed hierarchy tree $\mathbb{T}(\mathbb{W}, \Theta)$. In this formal

---

[11] The cardinality of a set $\mathbb{W}$ is denoted as $|\mathbb{W}| \in \mathbb{N}$.

[12] A minimalist example of a subspace extraction function is the establishment of the coordinates of a bounding box from the points describing the object model. Finer spatial resolution can be attained using more sophisticated methods such as *voxels* [47], *convex hull* [66], *octrees* [67] or *implicit surfaces* [68].

[13] The full containment is defined by the strict subspace, denoted as $\subsetneq$. This implies self-enclosing is not possible.

hierarchy, the object models O_i are connected to subordinate (fully contained) object models O_j in a lower hierarchy level, namely $O_j \succ O_i$. By observing the connectivity degree of the nodes in the spatial hierarchy tree, three different types of nodes are distinguished:

- **Root node**: There is one single root node $O_{\mathbb{W}}$ which is an abstract entity embracing the complete spatial domain of the representation

$$O_i \succ O_{\mathbb{W}} \ : \ \forall O_i \in \mathbb{W}. \tag{3.4}$$

- **Leaf nodes**: There is at least one leaf node O_l which encloses no subordinate object models

$$O_i \succ O_l \ : \ \nexists \ O_i \in \mathbb{W}. \tag{3.5}$$

- **Inner branch nodes**: Finally, there are nodes O_b which are neither leafs nor the root $O_{\mathbb{W}}$ node. They enclose at least one subordinate object model

$$O_i \succ O_b \ : \ \exists \ O_i \in \mathbb{W}. \tag{3.6}$$

Figure 3.11 represents an exemplary directed hierarchy tree. At the upper left corner, the subspace $\Theta(O_{\mathbb{W}})$ contained by the root node $O_{\mathbb{W}}$ delimits the whole spatial domain of the world model representation $\mathbb{W}$. Due to the acyclic structure of the hierarchy tree, it is possible to add, remove or even change the composition of an object model without affecting other spatially unrelated object models. This is the structural key for scalability.

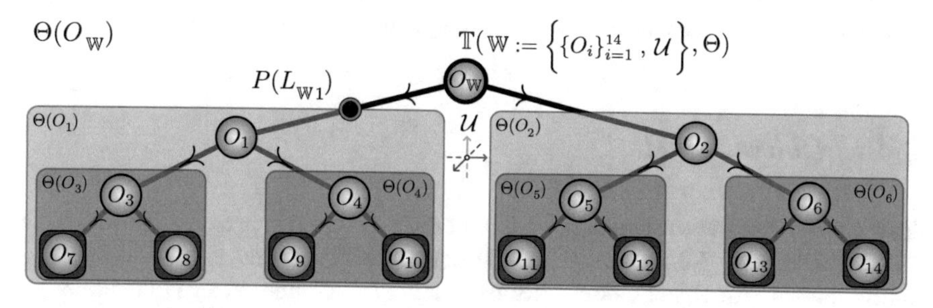

Fig. 3.11 The directed hierarchy tree $\mathbb{T}(\mathbb{W}, \Theta)$ from an exemplary world model representation $\mathbb{W}$ consisting of $n = 14$ object models. The hierarchy tree is topologically organized by the spatial enclosing principle (stated in Eqs. 3.2 and 3.3). The subspace contained by each model object is depicted as a gray rectangle. An instance of a parametric transformation $P(L_{\mathbb{W}1})$ (Eq. 3.7) shows a subtree which can be dynamically modified by parameters

3.4.3 Spatial Transformations

Depending on the object model and the application, a directed link of the hierarchy tree $\mathcal{L}_{ij} := (O_i, O_j) \Longleftrightarrow (O_j \succ O_i)$ connecting two nodes can contain a parametric transformation. An instance of such a dynamic transformation mechanism is a rigid transformation describing mostly parametric rotations, translations, or combinations of both. The dynamic mechanism can be expressed with

$$P(\mathcal{L}_{ij}) = \begin{cases} \mathcal{T}(T, \alpha, \beta, \theta) \in SE^3 \subset \mathbb{R}^{4x4}, & \text{if 6D-Transformation}: \\ & T\text{-Translation vector and } \alpha, \beta, \theta\text{-angles} \\ \mathcal{R} \in SO^3, \alpha, \beta, \theta \in \mathbb{R} & \text{if 3D-General-Rotation:} \\ & \alpha, \beta, \theta\text{-angles} \\ L \in \mathbb{R}^6, \omega \in \mathbb{R}, & \text{if 3D-Axis-Angle-Rotation:} \\ & L\text{-Axis and } \omega\text{-angle} \\ T \in \mathbb{R}^3, & \text{if 3D-Translation:} \\ & T\text{-Translation vector} \\ \emptyset, & \text{else.} \end{cases} \tag{3.7}$$

The connected collection of dynamic transformations $P(\mathcal{L}_{ij})$ within the directed hierarchy tree $\mathbb{T}(\mathbb{W}, \Theta)$ describes the kinematic tree $\mathbb{T}_P(\mathbb{W}, \Theta)$. Dynamic changes in the kinematic tree transform the positions and orientations of the object models. For instance, the subtree rooted at O_1 (see Fig. 3.11). These dynamic modifications would generate inconsistencies in the spatial hierarchy changing the containment of the object models. In order to avoid these inconsistencies and to hold the spatial hierarchy coherence, the spatial enclosing order is determined at the *initial-state*[14] of the world model representation. Hence, the spatial hierarchy is not modified by dynamic kinematic transformations in subsequent world model states.

3.4.4 Object Model

The world model representation $\mathbb{W}$ is an object-centered representation based on CAGM (see Sect. 3.3.7). Therefore, the boundary description is used to describe object models. An object model is formally a graph[15] composed as a four-tuple

$$O_i := \Big(V, E, F, C\Big) \in \mathbb{W}, \tag{3.8}$$

[14]The *world model state* is the particular configuration of all values of the parametric transformations of the world model (in Eq. 3.7) at certain time.

[15]The proposed definition of *object-graph* is equivalent to the standard mathematical graph by considering all vertices, edges, triangular faces and composed faces as attribute nodes, whereas the arcs interconnecting them are defined in the incidence list within each of theses attribute nodes.

where V stands for the *ordered set*[16] (a list) of vertices, E represents the edge set, F denotes the triangular face set and C stands for the composed face set. In the following sections, a detailed description of these elements is introduced followed by the description of their collective incidence structure.

Vertices

The points in 3D space describing the metric structure of an object model are called vertices $v_k \in \mathbb{R}^3$. The ordered list of m vertices constituting the object model O_i is denoted as

$$V(O_i) := \left\{ v_k \right\}_{k=1}^{m}. \tag{3.9}$$

The metric information of an object model (such as size) is implicitly contained within this vertex list. The object model metrics are decoupled from its topologic structure. This can be easily illustrated by a transformation affecting only the location of the vertices. For instance, a non-isotropic scaling transformation $\psi \in \mathbb{R}^{3x3}$ expressed as

$$\psi(v_k) = \acute{v}_k = diag\{s, s^2, s^3\}\ v_k,\ \ s > 1, s \in \mathbb{R}, \tag{3.10}$$

affects the size and ratios of the object model but does not change the topologic structure defined by the following shape-primitives.

Edges

CAGM representations are polyhedral approximations of the object shapes. In these representations, line segments joining adjacent vertices are called geometric edges. Hence, edges arise at the intersection of two surfaces. This means according to [47], the CAGM representation is circumscribed to homogeneous 2D topology polyhedra. Formally, an edge is defined as

$$e_{\alpha\beta} := (v_\alpha, v_\beta)\ \mid\ v_\alpha, v_\beta \in V(O_i) \Rightarrow \alpha, \beta \in \mathbb{N}^+ \text{, where } (1 \le \alpha < \beta \le m). \tag{3.11}$$

The edge $e_{\alpha\beta}$ connects the vertex v_α to the vertex v_β. Its length is determined by the l^2 norm of the translation vector

$$\Gamma(e_{\alpha\beta}) := ||v_\beta - v_\alpha|| \in \mathbb{R}^+. \tag{3.12}$$

Due to the vertex ordering ($\alpha < \beta$), there is no ambiguity (neither loops nor multi-graphs) in the edge composition. Consequently, the object model O_i contains an edge set $E(O_i)$ expressed as

$$E(O_i) := \left\{ e_{\alpha\beta} \right\}_{\alpha,\beta \in V(O_i)}^{p} \subset \left\{ v_\alpha \otimes v_\beta\ |\alpha < \beta \right\}, \tag{3.13}$$

[16]This ordering is required for the description of shape-primitives such as edges, triangular faces and composed faces.

where $\otimes$ is the Cartesian product. Thus, the cardinality p of the edge set

$$p \leq \frac{m(m-1)}{2} \tag{3.14}$$

is upper bounded by the vertex list cardinality $m = |V(O_i)|$. Now that the vertices and edges are defined, it is possible to introduce the valence of a vertex $\nu(v_\chi) : v_\chi \in V(O_i) \mapsto \mathbb{N}^+$ as the amount of edges connected to v_χ,

$$\nu(v_\chi) := \Big|\{e_{\alpha\beta} \mid (\chi = \alpha) \vee (\chi = \beta)\} \subset E(O_i)\Big|. \tag{3.15}$$

The valence of a vertex reflects its connectivity degree and is important when applying filtering procedures in composed searches (see Sect. 3.4.5).

Triangular Faces

The description of general smooth surfaces is approximated by surface subdivision. Depending on the application, this process can be realized with methods (see examples in [69, 70]). Independently of the selected method, the result is a net of planar *polygonal*[17] oriented faces forming closed sequences. In the case of constructive solid geometric (boolean composition of polyhedron shapes), the polygonal faces naturally describe the surfaces in an explicit and computationally accessible manner (see examples in Sect. 3.3.7). Due to the simplicity of a triangle net (which can be described with linear equations) translated into computational advantages,[18] most common polygonal surfaces used in CAGM representations are triangular faces. A triangular face denoted as $f_{\alpha\beta\chi}$ is uniquely defined by three non-collinear vertices v_α, v_β and v_χ holding the order constraint $(\alpha < \beta) \wedge (\alpha < \chi) \wedge (\beta \neq \chi)$. The indices order β and χ is important to be unrestricted because it determines which of the two possible face-orientations is intended. The face orientation is expressed by the unitary normal vector $\hat{N}$ from

$$N(f_{\alpha\beta\chi}) = (v_\beta - v_\alpha) \times (v_\chi - v_\alpha) \in \mathbb{R}^3, \tag{3.16}$$

where $\times$ stands for the cross product. In these terms, the area of a triangular face $A(f_{\alpha\beta\chi})$ is expressed as

$$A(f_{\alpha\beta\chi}) = \frac{1}{2}||N(f_{\alpha\beta\chi})|| \in \mathbb{R}^+. \tag{3.17}$$

Finally, the union of all elements contained in the triangular face set

---

[17] A planar polygon is defined as *"a closed plane figure bounded by three or more line segments that terminate in pairs at the same number of vertices and do not intersect other than at their vertices"*. Extracted from [71].

[18] For example, convex structure, coplanarity, easy accessibility, verbosity, etc.

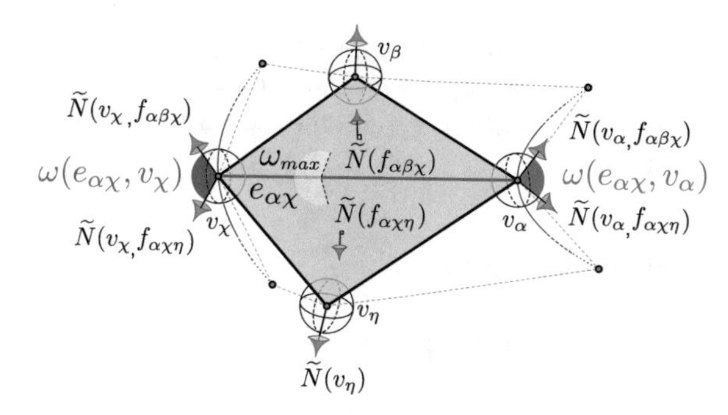

Fig. 3.12 The maximal angle of aperture of the edge $\omega_{max}(e_{\alpha\beta}, v_\alpha)$ (from the subdivision of the smooth surface in blue lines) is expressed as the maximal angle of aperture between the normals involved in the incidence on both triangular faces $f_{\alpha\beta\chi}$ and $f_{\alpha\chi\eta}$. This property is correlated with the edge visibility or saliency

$$F(O_i) := \left\{ f_{\alpha\beta\chi} \right\}^q_{\alpha,\beta,\chi \in V(O_i)}, \tag{3.18}$$

defines the boundary of the object model O_i. Its cardinality q is upper bounded by the binomial coefficient $q < \binom{p}{k=3}$, where $p = |E(O_i)|$ stands for the cardinality of the edge set (Eq. 3.14) and $k = 3$ denotes the three edges of the triangular faces.

Now, it is possible to present a concept involving both edges and faces. The angle of aperture ω of an edge $e_{\alpha\beta}$ at the vertex v_α is the angular relation

$$\omega(e_{\alpha\beta}, v_\alpha) = \arccos\left(\widetilde{N}(v_\alpha, f_{\alpha\beta\chi}) \cdot \widetilde{N}(v_\alpha, f_{\alpha\chi\eta})\right), \tag{3.19}$$

where $\widetilde{N}(v_\alpha, f_{\alpha\beta\chi})$ is the normal vector associated to the vertex v_α when defining smooth surfaces by triangular approximation (see Sect. 3.4.6). The edge angle of aperture is bounded as $0 \leq \omega(e_{\alpha\beta}, v_\alpha) \leq \pi$. The maximal edge angle of aperture is defined by considering both incident triangular faces (see Fig. 3.12) as

$$\omega_{max}(e_{\alpha\beta}) := \mathbf{argmax}_{v_i \in \{v_\alpha, v_\beta\}}\, \omega(e_{\alpha\beta}, v_i). \tag{3.20}$$

Although the use of triangular faces $f_{\alpha\beta\chi} \in F(O_i)$ have computational advantages, there are higher discriminant polygonal faces. For instance, common and easy to recognize environmental elements, such as rectangular surfaces in doors, tables, etc. (see segmentation based on rules in [63]) can be useful for visual recognition. The subdivision of such polygonal faces into triangular sub-faces generates *virtual edges*.[19] Despite their continuous physical composition, the area of these rectangular surfaces is only implicitly-stored within the world model representation. These two issues,

[19] These edges are auxiliary elements used to describe complex surfaces by simpler triangular nets in order to provide computational advantages.

virtual edges and implicit-stored attributes, avoid the proper world model indexing for object recognition using correspondence matching methods. These issues were avoided in the proposed representation with the help of the visibility attribute and composed face entity.

Visibility

The binary visibility attribute for edges

$$\Delta(e_{\alpha\beta}) \ : \ E(O_i) \mapsto \{0, 1\} \tag{3.21}$$

and faces

$$\Delta(f_{\alpha\beta\chi}) \ : \ F(O_i) \mapsto \{0, 1\}, \tag{3.22}$$

are introduced. These attributes are used to express whether the elements can physically manifest and be observed[20] on images from at least one reachable viewpoint. Based on these attributes, the amount of visible edges connected to the vertex v_k is called *non-virtual valence* $\hat{\nu}(v_k) \mapsto \mathbb{N}^+$ and it expressed (as an extension of Eq. 3.15), namely

$$\hat{\nu}(v_k) := \left|\left\{ e_{\alpha\beta} \mid \Big((k = \alpha) \vee (k = \beta) \Big) \wedge \Big(\Delta(e_{\alpha\beta}) = 1 \Big) \wedge \Big(e_{\alpha\beta} \in E(O_i) \Big) \right\}\right|. \tag{3.23}$$

The non-virtual valence is useful for querying and recognition by geometric incidence expansion. In Sect. 3.4.5, these attributes can be used to dismiss invisible shape-primitives. The detection and removal of indiscernible elements improves the processing performance and reduces the storage space of the world model representation.

Composed Faces

A composed face is defined as a collection of coplanar triangular faces sharing a common orientation and adjacent edges. Formally, a composed face is expressed as

$$S_h := \Big\{ f_{\alpha\beta\chi} \Big\}_{[\alpha,\beta,\chi] \in V(O_i)} \subset F(O_i), \tag{3.24}$$

where all triangular faces are subject to common orientation

$$\forall f_{\alpha\beta\chi}, f_{\eta\iota\kappa} \in S_h \Leftrightarrow \widetilde{N}(f_{\alpha\beta\chi}) \cdot \widetilde{N}(f_{\eta\iota\kappa}) > \epsilon \tag{3.25}$$

where $(\cdot)$ denotes the scalar or inner product and $\epsilon \approx 1$ is a threshold which considers numeric errors of the floating-point representation. The common non-visible edge between two triangular faces is expressed as $\forall f_{\alpha\beta\chi} \in S_h \ \exists \ f_{\eta\iota\kappa} \in S_h$ then

$$\left| \Big\{ (\alpha, \beta), (\alpha, \chi), (\beta, \chi) \Big\} \cap \Big\{ (\eta, \iota), (\eta, \kappa), (\iota, \kappa) \Big\} \right| = 1, \tag{3.26}$$

[20]Manifestable properties are presented in Sects. 3.4.6 and 3.5.

In addition, all triangular faces $f_{\alpha\beta\chi}$ contained in a multiple face S_h store a reference to the composed face where they belong $K : F(O_i) \mapsto \mathbb{N}$, namely

$$K(f_{\alpha\beta\chi}) = h \Leftrightarrow f_{\alpha\beta\chi} \in S_h. \quad (3.27)$$

This reference is necessary during recognition and matching validation. For example, when using the area as a filtering cue. Hence, the area of a composed face is expressed in terms of Eq. 3.17 as

$$A(S_h) := \sum_{\alpha,\beta,\chi \in S_h}^{r} A(f_{\alpha\beta\chi}). \quad (3.28)$$

Finally, the set of composed faces of an object model O_i is denoted as

$$C(O_i) := \{S_h\}_{\mathrm{H}}. \quad (3.29)$$

This can also be an empty set (see Fig. 3.13). Finally, the definition of a composed face S_h does not imply a convex shape as in the case of the triangular faces (see Fig. 3.30).

Redundancy

An important observation about the composed face entity is the following fact. All edges $e_{\tau\upsilon}$ whose incident triangular faces $e_{\tau\upsilon} \in f_{\alpha\beta\chi}$ and $e_{\tau\upsilon} \in f_{\eta\iota\kappa}$ are inside a composed face S_h are virtual edges and are denoted as

$$(e_{\tau\upsilon} \in f_{\alpha\beta\chi}) \wedge (e_{\tau\upsilon} \in f_{\eta\iota\kappa}) \wedge (f_{\alpha\beta\chi}, f_{\eta\iota\kappa} \in S_h) \Rightarrow \breve{e}_{\tau\upsilon}. \quad (3.30)$$

The direct consequence is that all virtual edges are non-visible

$$\Delta(\breve{e}_{\tau\upsilon}) = 0, \ \forall \breve{e}_{\tau\upsilon} \in E(O_i). \quad (3.31)$$

Incidence Structure

Because of the topologic cycles of the boundary description, an object model is a graph (Eq. 3.8). The vertices, edges, triangular faces and composed faces are arranged and recurrently linked to each other by an inherent incidence function

$$\Pi(O_i) : \left\{ V \cup E \cup F \cup C \right\} \mapsto \left\{ V \cup E \cup F \cup C \right\}. \quad (3.32)$$

In this work, there are structural differences to previous graph-based world model representations. For instance, in the attributed adjacency graph (AAG) used for object-representation in Sect. 3.3.8, the arcs of the graph store attributes which are specially selected for a particular detection, for example, the convex and concave edges (in Fig. 3.8). In contrast, the proposed graph representation in this work, the arcs of the graph are pure geometric adjacencies. These arcs do not hold additional struc-

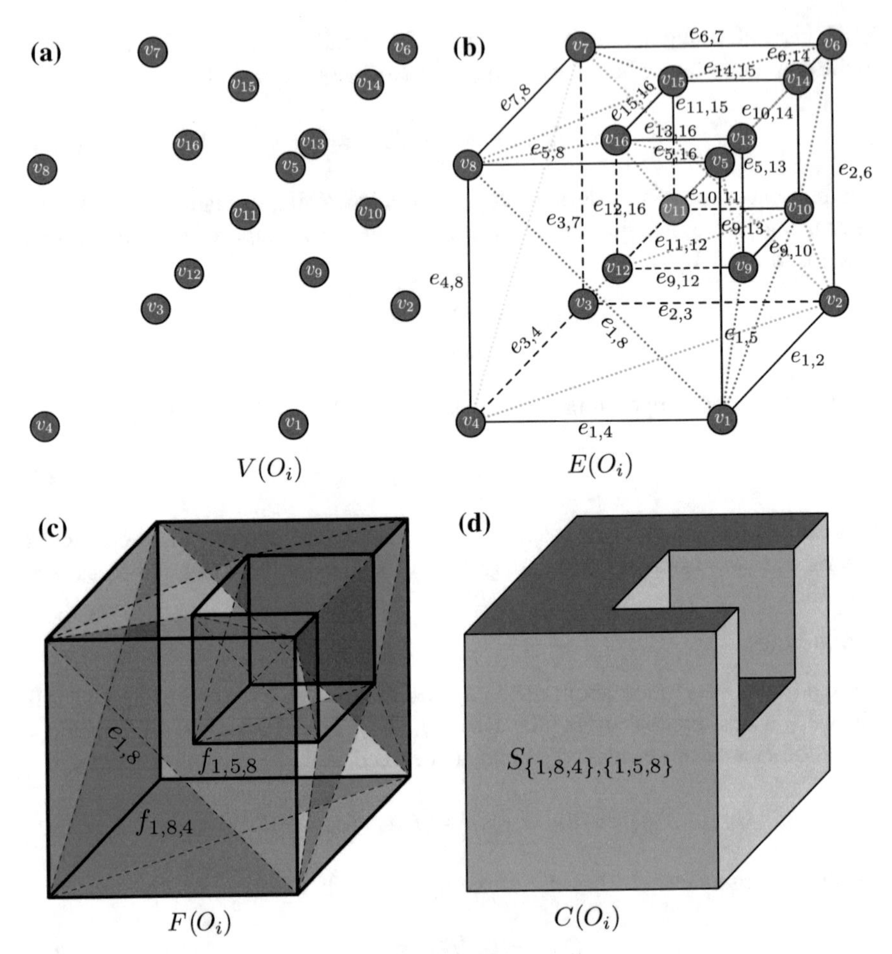

Fig. 3.13 Element-wise decomposition of an object model. The object model is a graph $O_i := (V, E, F, C)$ consisting of one list and three sets: (i) A vertex list $V(O_i)$ (Eq. 3.9). (ii) An edge set $E(O_i)$ (Eq. 3.13). (iii) A triangular face set $F(O_i)$ (Eq. 3.18). (iv) A composed faces set $C(O_i)$ (Eq. 3.29). **a** The vertex list $V(O_i)$ contains metric information of the object model. **b** The triangular face set $F(O_i)$ holds the topologic structure simultaneously defining the surface boundary and edge connectivity. Notice the non-occluded (from this particular viewpoint) visible edges are shown in continuous-black lines, whereas the still visible but occluded edges are displayed in dashed-black lines. The virtual edges $\breve{e}_{\tau\upsilon}$ are (as expected non-visible $\Delta(\breve{e}_{\tau\upsilon}) = 0$, see Eq. 3.31) shown by dotted-colored lines. **c** The triangular face set is shown with transparency for better visualization. **d** The composed face $S_{\{1,8,4\},\{1,5,8\}}$

tural attributes. This has been done for the purpose of keeping generality for wider applicability with diverse and complementary sensing modalities. This design guideline has no detrimental performance or robustness effects for searching or matching shape-primitives (see Sect. 3.4.5 for more details). In order to study and exploit the incidence structure of the object models, the conveniently pairwise decomposition of the incidence function $\Pi(O_i)$ is presented as follows:

Vertices ⟷ Edges

The structural composition of the edges implies the *surjective*[21] function $\Psi : (V(O_i), V(O_i)) \mapsto E(O_i)$ between pairs of vertices and edges

$$\Psi(v_\alpha, v_\beta) = \begin{cases} e_{\alpha\beta}, & \text{if } (\alpha < \beta) \wedge (e_{\alpha\beta} \in E(O_i)) \\ e_{\beta\alpha}, & \text{if } (\beta < \alpha) \wedge (e_{\beta\alpha} \in E(O_i)) \\ \emptyset, & \text{else} \end{cases} \tag{3.33}$$

with a *right-inverse*[21] function denoted as $\widetilde{\Psi} : E(O_i) \mapsto (V(O_i), V(O_i))$

$$\widetilde{\Psi}(e_{\alpha\beta}) = \begin{cases} (v_\alpha, v_\beta), & \text{if } (\alpha < \beta) \wedge (e_{\alpha\beta} \in E(O_i)) \\ (v_\beta, v_\alpha), & \text{if } (\beta < \alpha) \wedge (e_{\beta\alpha} \in E(O_i)) \\ \emptyset, & \text{else.} \end{cases} \tag{3.34}$$

.

Vertices ⟷ Faces

The triangular face composition is described by three non-collinear ordered vertices or through its boundary edges. By considering the incidence of the vertices, the surjective mapping function from vertex triplets to triangular faces $\Phi : (V(O_i), V(O_i), V(O_i)) \mapsto F(O_i)$ is expressed as

$$\Phi(v_\alpha, v_\beta, v_\chi) = \begin{cases} f_{\alpha\beta\chi}, & \text{if } (\alpha < \beta) \wedge (\alpha < \chi) \wedge (\beta \neq \chi) \wedge (f_{\alpha\beta\chi} \in F(O_i)) \\ f_{\alpha\chi\beta}, & \text{if } (\alpha < \beta) \wedge (\alpha < \chi) \wedge (\beta \neq \chi) \wedge (f_{\alpha\chi\beta} \in F(O_i)) \\ \emptyset, & \text{else.} \end{cases} \tag{3.35}$$

Analog to Eq. 3.34, the right-inverse function mapping from triangular faces to vertex triplets $\widetilde{\Phi} : F(O_i) \mapsto (V(O_i), V(O_i), V(O_i))$ is

$$\widetilde{\Phi}(f_{\alpha\beta\chi}) = \begin{cases} (v_\alpha, v_\beta, v_\chi), & \text{if } (\alpha < \beta) \wedge (\alpha < \chi) \wedge (\beta \neq \chi) \wedge (f_{\alpha\beta\chi} \in F(O_i)) \\ (v_\alpha, v_\chi, v_\beta), & \text{if } (\alpha < \beta) \wedge (\alpha < \chi) \wedge (\beta \neq \chi) \wedge (f_{\alpha\chi\beta} \in F(O_i)) \\ \emptyset, & \text{else.} \end{cases} \tag{3.36}$$

[21]A function is called surjection or *"onto"* if every element of its codomain has a corresponding element in its domain and multiple elements of its domain might be mapped *onto* the same element of the codomain. The inverse mapping of a surjection is called *right-inverse*.

Edges ⟷ Faces

By considering the incidence of edges and triangular faces, the surjective mapping from edge triplets to faces $\Lambda : (E(O_i), E(O_i), E(O_i)) \mapsto F(O_i)$ is expressed as

$$\Lambda(e_{\alpha\beta}, e_{\alpha\chi}, e_{\beta\chi}) = \begin{cases} f_{\alpha\beta\chi}, & \text{if } (\alpha < \beta) \wedge (\alpha < \chi) \wedge (\beta \neq \chi) \wedge (f_{\alpha\beta\chi} \in F(O_i)) \\ f_{\alpha\beta\chi}, & \text{if } (\alpha < \beta) \wedge (\alpha < \chi) \wedge (\beta \neq \chi) \wedge (f_{\alpha\chi\beta} \in F(O_i)) \\ \emptyset, & \text{else.} \end{cases} \tag{3.37}$$

In the same manner (as Eqs. 3.34 and 3.36), the right-inverse maps from faces to edge triplets $\widetilde{\Lambda} : F(O_i) \mapsto (E(O_i), E(O_i), E(O_i))$ is expressed as

$$\widetilde{\Lambda}(f_{\alpha\beta\chi}) = \begin{cases} e_{\alpha\beta}, e_{\alpha\chi}, e_{\beta\chi}, & \text{if } (\alpha < \beta) \wedge (\alpha < \chi) \wedge (\beta \neq \chi) \wedge (f_{\alpha\beta\chi} \in F(O_i)) \\ e_{\alpha\chi}, e_{\alpha\beta}, e_{\chi\beta}, & \text{if } (\alpha < \beta) \wedge (\alpha < \chi) \wedge (\beta \neq \chi) \wedge (f_{\alpha\chi\beta} \in F(O_i)) \\ \emptyset, & \text{else.} \end{cases} \tag{3.38}$$

The partial mapping functions $\Psi(O_i)$, $\Phi(O_i)$ and $\Lambda(O_i)$ summarize the global mapping function $\Pi(O_i)$ (see Fig. 3.14). The pairwise decomposition enables a compact description of algorithms within this work. This formulation clearly unveils the fact that object models O_i are tripartite graphs. This occurs because $V(O_i)$, $E(O_i)$ and $F(O_i)$ are disjoint sets, namely their elements are not directly adjacent (see Fig. 3.15). This means, when finding the connection from one vertex to another, the particular connecting edge or one of the two possible incident triangular faces is required. Likewise, in order to establish the link between two edges, the common vertex or triangular face must exist in the object model. In the same manner, the incidence of two triangular faces is determined by either common vertex or edge. This linkage is of critical importance for recognition and validation using correspondence methods as in the interpretation tree [24]. Finally, the set of composed faces $C(O_i)$ is not considered directly in the partial incidence mapping. This occurs because this is not a fundamental set but an agglutination, namely a derived subset of $F(O_i)$. Because of the bidirectional linkage of edges, faces and composed faces, the set $C(O_i)$ has an important role for recognition. Due to this bidirectional linkage, it is possible to efficiently extract the composed faces (see Sect. 3.4.5).

3.4.5 Probabilistic Indexing

The world model representation of this work includes efficient mechanisms for information retrieval. The information to be recalled consists of geometrically structured subsets of shape-primitives. The mechanisms deriving this information retrieval are based on the exploitation of scalar attributes of the shape-primitives. This association is done by efficient search queries. These foundations have been inspired by the methods in [72–74].

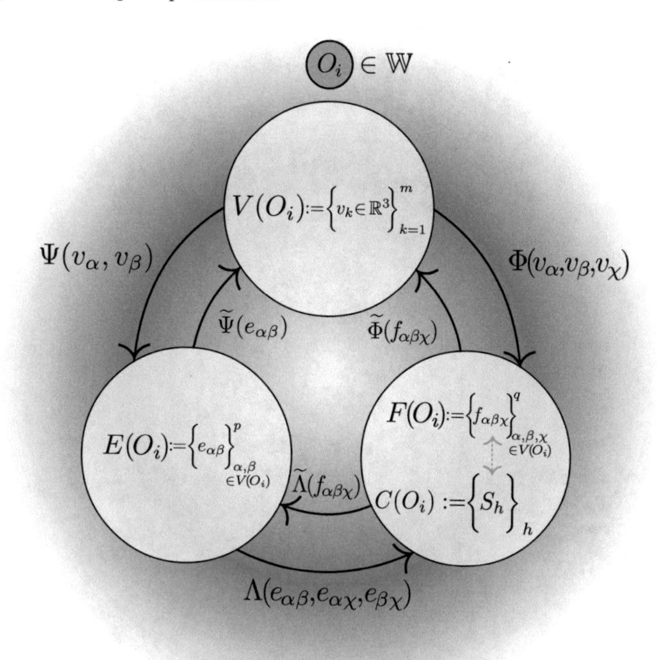

Fig. 3.14 The tripartite graph structure of the object model $O_i := (V, E, F, C)$. The incidence structure of the object models is described by the incidence functions Ψ, Φ, Λ and their *right-inverse* counterparts (from Eq. 3.33 until Eq. 3.38). The composed face set $C(O_i)$ is linked only to the triangular face set $F(O_i)$. This mapping is stored in the attribute of the triangular face $f_{\alpha\beta\chi} \in S_h$ as $K(f_{\alpha\beta\chi})$ (Eq. 3.27)

The attributes used for indexing object models are scalar magnitudes which can be systematically organized according to their relevance within a particular task. The concept of relevance is context-dependent and it varies from one search query to another (see the peek-effect in [72]). Additionally, within a particular search query, the relevance of an element has a likelihood interpretation which leads to a coherent formulation for filtering and matching in probabilistic robot perception.

Due to its object-centered nature, the indexing schema is viewpoint independent in the world model representation of this work. This is a fundamental difference compared to other approaches which index different combinations of shape-primitives according to projective or affine transformations. Examples of those approaches are the projective consistency constraints (see Sect. 3.3.4), the visual alignment (presented in Sect. 3.3.5) or geometric hashing (see [26]). In order to provide a sequential formulation of the proposed indexing schema, the formal definition of the object model indexing is presented (in Sect. 3.4.5). Afterwards, the probabilistic aspects involved in the search procedures are introduced (in Sect. 3.4.5).

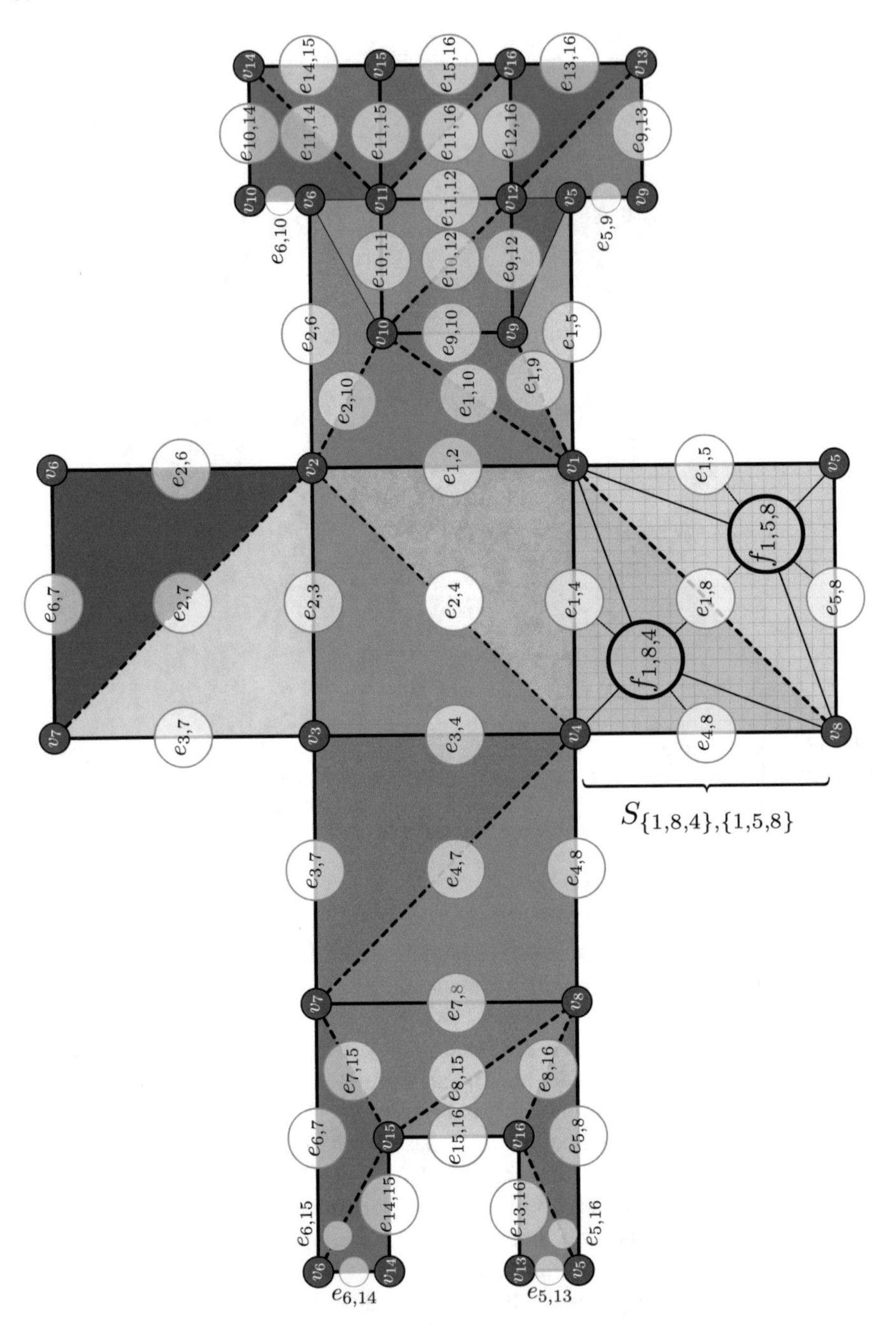

Fig. 3.15 The incidence graph functions of an object model are explicitly stored in the composition of the shape-primitives. This planar representation (from Fig. 3.13) shows the virtual edges (displayed by dotted-lines, for instance $e_{1,8}$) and composed faces (such as $S_{\{1,8,4\},\{1,5,8\}}$). This visualization provides an insight into the storage and recalling complexity of a rather simple object. Notice that a few nodes of the graph are repeated to embed the object model in the plane, for example at the vertices v_{15} and v_{16} or the edges $e_{6,10}$ and $e_{5,9}$

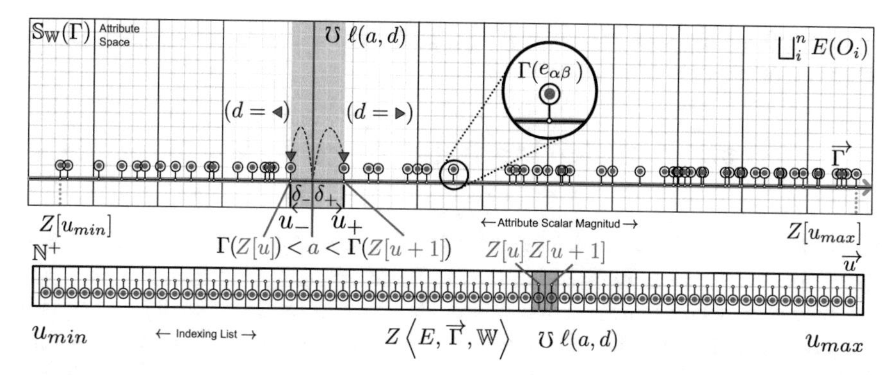

Fig. 3.16 The attribute-based indexing search. This example shows visible edges indexed by increasing length order. The upper part of the figure shows the continuous attribute space $\mathbb{S}_{\mathbb{W}}(\Gamma)$ (notice the single sample in the zoom-circle). The lower part of the figure shows the indexing list Z where the elements are consecutively located in a discrete space. The indexing search $\mho$ is marked with a gray dark region in both spaces. The selection criterion ℓ is displayed with both selection directions ($\triangleright$, $\triangleleft$) and corresponding length deviations $\delta(u_{\pm})$

Attribute-Based Indexing

The search of a particular shape-primitive is realized by filtering scalar attributes fulfilling metric criteria. This mechanism can be efficiently computed by organizing the shape-primitives in a monotonic progression. The scalar attributes vary from one shape-primitive to another. For example, the length of an edge $\Gamma(e_{\alpha\beta})$ (Eq. 3.12), the area of a face $A(f_{\alpha\beta\chi})$ (Eq. 3.17), the valency of a vertex $\nu(v_k)$ (Eq. 3.15), etc. (see Fig. 3.16). Additionally, the attributes differ from one query procedure to another. For instance, searching for a visible edge while detecting a door handle or searching for a composed face based on the perceived area of a rectangular surface. An important example of attribute-based indexing is the length $\Gamma(e_{\alpha\beta})$ of the visible edges $\Delta(e_{\alpha\beta}) = 1 \mid \forall e_{\alpha\beta} \in \mathbb{W}$. Its importance relies on the fact that length is a very discriminant cue for visual recognition, especially when using calibrated stereo vision. Therefore, the example of edge length as indexing attribute is successively used along this chapter in order to facilitate the derivation of formal definitions.

Indexing List

The edge length indexing requires a sorted list[22] storing all numerical references (from now on indices) to the edges existing in all object models within the world model representation. The indexing list $Z : u \in \mathbb{N}^+ \mapsto e_{\alpha\beta}$ is an ordered set. Its comprehensible arrangement is determined by a predicate function which compares the indexing attribute from the shape-primitive. In the edge length example, this

[22] Associative arrays (such as hash tables) were not considered due to: (i) The multiple hashing collisions occurring in symmetric elements. (ii) The fact that the estimation and querying of the PDF in the attribute space (see Sect. 3.4.5) using other data structure (rather than a list) is less efficient.

predicate function is the function Γ (Eq. 3.12). The indexing arrangement of the edge length example is expressed as

$$Z\left\langle E, \overrightarrow{\Gamma}, \mathbb{W}\right\rangle = \bigcup_{i}^{n} E(O_i) \quad | \; \Gamma(Z[u]) \leq \Gamma(Z[u+1]), \tag{3.39}$$

where the angular brackets are used for the definition of the indexing list. Next, the direction of the arrow on top of the predicate function $\overrightarrow{\Gamma}$ defines the predicate increasing or decreasing direction of the elements in the list. Finally, the square brackets $Z[u]$ stand for the reference (also known as *"address of"*) operator within the bounded domain $(u_{min} := 1) \leq u \leq (u_{max} := \sum_i^n |E(O_i)|)$. Because the indexing list contains all[23] visible and salient edges of the world model, only the indices[24] are stored in order to avoid unnecessary memory usage and processing overhead.

Indexing Search

The search of a shape-primitive in an indexing list is an inexact predicate comparison which is *optimally*[25] computed using a binary search. The selection criterion steering such binary search is a two-tuple ℓ consisting of the scalar perceptual cue a and the progression direction d, namely

$$\ell(a \in \mathbb{R}, d \in \{\triangleleft, \triangleright\}). \tag{3.40}$$

The value a should lie within the sorting attribute boundaries of the indexing list $\Gamma(Z[u_{min}]) \leq a \leq \Gamma(Z[u_{max}])$. The direction parameter d is necessary to determine the first element in the progression which is either ($\triangleright$) bigger or ($\triangleleft$) smaller than the scalar value a. This parameter takes into account the cases where the selection value a lies between two elements in the indexing list, namely $\Gamma(Z[u]) < a < \Gamma(Z[u+1])$. In such situations, the direction parameter determines which is the first index fulfilling the criterion $\ell(a, \triangleleft) \mapsto u$ or well $\ell(a, \triangleright) \mapsto u+1$. An indexing search $\mho : \ell(a, d) \mapsto \{(u_{min} \leq u \leq u_{max}) \in \mathbb{N}^{+} \cup \emptyset\}$ is uniquely parametrized by the selection criterion ℓ. An indexing search is expressed using the indexing list (Eq. 3.39)

$$\mho\left\{\ell(a, d), Z\right\} := \begin{cases} (\emptyset, 0), & \text{if } (a < \Gamma(Z[u_{min}])) \vee (a > \Gamma(Z[u_{max}])) \\ (u, 0) & \text{if } \Gamma(Z[u]) \equiv a \\ (u_{+}, \delta_{+}) & \text{if } (\Gamma\,(Z[u_{+}]) > a) \wedge (d = \triangleright) \\ (u_{-}, \delta_{-}), & \text{if } (\Gamma\,(Z[u_{-}]) < a) \wedge (d = \triangleleft), \end{cases} \tag{3.41}$$

[23] See discussion about virtual and non-salient edges in Sect. 3.4.6.

[24] The integer value corresponding to the memory allocations were the instance is stored. This is actually inspired by the *in-memory database* systems [75].

[25] The worst performance case is of order $O(log_2(u_{max}))$.

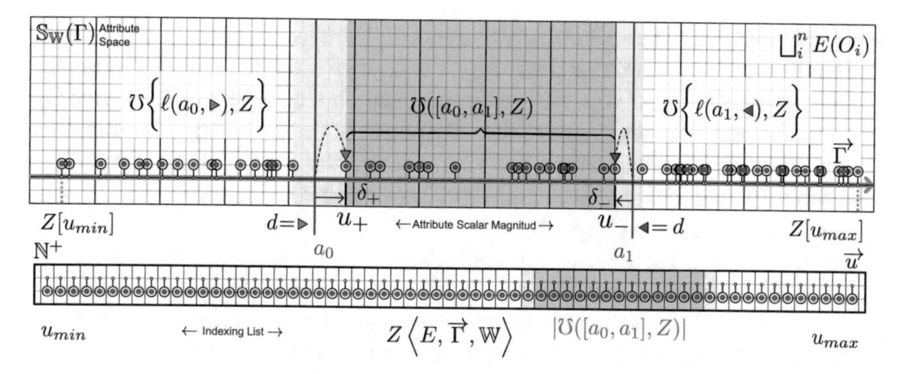

Fig. 3.17 The interval search $\mho([a_0, a_1], Z)$ determines shape-primitives fulfilling an interval criterion. This figure shows edge length indexing and the value pair $[a_0, a_1]$. The interval search is the basis for probabilistic queries for coupling visual-features with shape-primitives of the world model representation (see Sect. 3.4.5)

where u_+ and u_- denote the indices satisfying the criterion $\ell(a, d)$ with a positive or negative distance to the selection value a, namely $\delta(u_\pm) = \Gamma(Z[u_\pm]) - a$. Figure 3.16 shows an indexing search (Eq. 3.41).

Interval Search

The concept of indexing search $\mho$ is extended to an interval search by defining two different attribute values a_0 and a_1 within the domain of the sorted list $\Gamma(Z[u_{min}]) \leq a_0 < a_1 \leq \Gamma(Z[u_{max}])$. The search interval is expressed (in terms of Eq. 3.41) as

$$\mho([a_0, a_1], Z) := \left[\mho\left\{\ell(a_0, \rhd), Z\right\}, \mho\left\{\ell(a_1, \lhd), Z\right\}\right], \tag{3.42}$$

with $0 \leq |\mho([a_0, a_1], Z)| \leq u_{max}$ elements (see Fig. 3.17).

Probabilistic Search Queries

The scalar attribute value of each shape-primitive extracted by a predicate function (for example $\Gamma(e_{\alpha\beta})$) is a sample point along the one dimensional continuous space called *attribute space* $\mathbb{S}_{\mathbb{W}}(\Gamma) \subset \mathbb{R}^+$. The indexing of these sample points provides an efficient mechanism to search shape-primitives either by single value or through a range of values. In addition, an adept shape retrieval mechanism must provide selection metrics based on likelihood. Therefore, the analysis of the distribution of the shape-primitives can provide the means to achieve search queries according to their relevance in terms of likelihood criteria. In this context, the relevance depends on whether the matching process (performing the search queries) requires to attain the most likely elements (expected cues) or the less frequent elements (discriminant cues) within the attribute space. According to the objective, the analysis can be categorically split into:

- **Global analysis**: The global analysis concerns with the probability density function (PDF) of the sample values $P : \mathbb{R}^+ \mapsto \mathbb{R}$. It provides the basis for estimating the density associated with any particular location along the attribute space $\mathbb{S}_{\mathbb{W}}$. The estimated PDF also enables the assessment of the probabilities within intervals defined by frontier values. The estimation of the density of shape-primitives according to particular attributes allows the composition of queries in a conditional probability sense (see Chap. 7).
- **Local analysis**: The distribution analysis of shape-primitives within a fraction of the attribute space provides the key to handle uncertainty considerations for noise prone selections, namely the management of uncertain perceptual cues (a's in Eq. 3.40). This is realized by locally determining elements found above certain confidence density. The density spread and profile of the uncertainty are determined according to the perceptual deviation from the visual uncertainty model (see details in Chap. 6).

Global-PDF Search Queries

Following the edge length example, the probability density function P_Γ is approximated by non-parametric kernel density estimation of the edge length $D_\Gamma(a)$ ([76]) as follows

$$P_\Gamma(a) \approx D_\Gamma(a) = \frac{1}{\sum_i^n |E(O_i)|} \sum_{j=1}^{\sum_i^n |E(O_i)|} \mathbf{exp}\left[\frac{[\Gamma(e_i) - a]^2}{2\varsigma^2}\right], \tag{3.43}$$

within the closed-domain $\Gamma(Z[u_{min}]) \leq a \leq \Gamma(Z[u_{max}])$ such as

$$P_\Gamma\Big([\Gamma(Z[u_{min}]), \Gamma(Z[u_{max}])]\Big) = 1, \tag{3.44}$$

where the 1D-Gaussian kernel is applied with bandwidth ς attained by the *Silverman* rule ([77]) as

$$\iota_p = \frac{1}{\sum_i^n |E(O_i)|} \sum_{j=1}^{\sum_i^n |E(O_i)|} \Gamma(e_j)^p, \tag{3.45}$$

$$\varsigma = \left[\frac{4}{3\sum_i^n |E(O_i)|} \underbrace{(\iota_2 - \iota_1^2)^{\frac{5}{2}}}_{\text{variance}}\right]^{\frac{1}{5}}, \tag{3.46}$$

where the p-raw moment ι_p allows a compact representation of the standard deviation (see the resulting PDF in Fig. 3.18).

The continuous density function D_Γ enables generalized estimations of the probability density along the attribute space (length of edges) even in intervals with no samples. The density estimation of each sample inside the indexing list Z allows direct and fast comparisons (see Fig. 3.19). The modifications of the world model

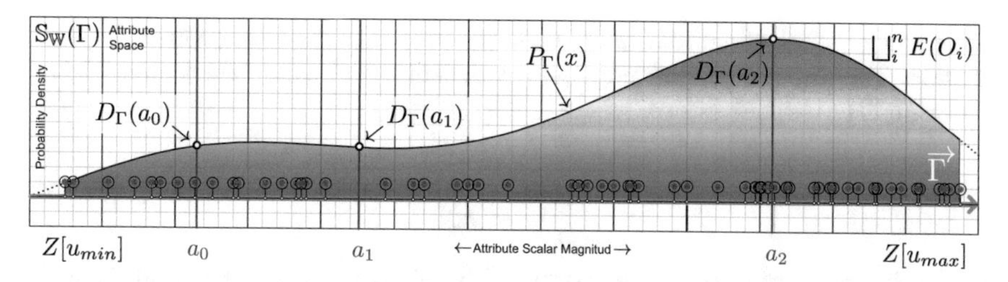

Fig. 3.18 The probability density function $P_\Gamma(a)$ is approximated by kernel density estimation $D_\Gamma(a)$ (Eq. 3.43). Notice the three different length selection criteria a_0, a_1 and a_2. The associated density of the first two locations is the same $D_\Gamma(a_0) = D_\Gamma(a_1)$ despite their length difference ($a_1 > a_0$) and the absence of samples directly under a_1. This singular situation shows the importance of the density analysis to minimize or maximize the probability of a search query. The global maximal density is found at a_2. At the lower left and right limits, the black dotted-lines show the tendency of the density distribution

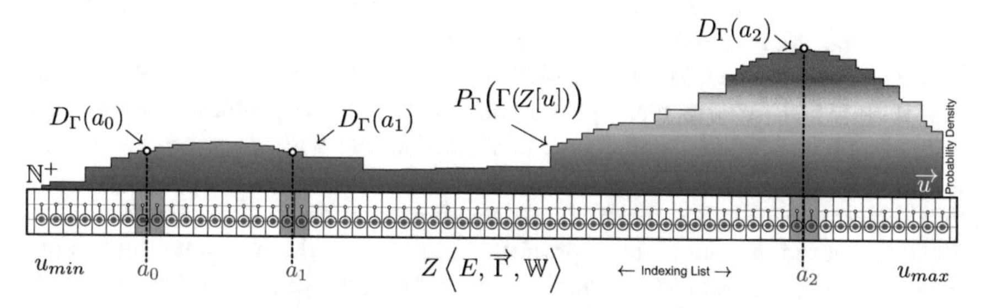

Fig. 3.19 The indexing list is also a look-up table of the density distribution reflecting the implicit PDF of the samples $D_\Gamma(\Gamma(Z[u]))$. Notice that the discrete profile of the look-up table is not a direct discretization of the samples (in Fig. 3.18). This occurs because of the irregular distribution of the edge length in the continuous domain $\mathbb{S}_{\mathbb{W}}(\Gamma)$

representation affecting the length or the amount of edges require the re-computation of this look-up table in order to coherently reflect the world model structure.

A direct use of the probability density function P_Γ for visual perception is the estimation of the probability of an interval. This likelihood metric enables the selective integration of perceived visual-features. This improves the computation performance while creating the association between visual-features and shape-primitives. This reduces the amount of combinations to be considered in order to attain the matching. It also maximizes the probability of a successful outcome. The continuous approximation of the interval probability is denoted as

$$P_\Gamma\left([a_0, a_1]\right) \approx \frac{1}{\epsilon_d} \sum_{u=\mho\{\ell(a_0,\triangleright), Z\}}^{\mho\{\ell(a_1,\triangleleft), Z\}} D_\Gamma\bigg(Z[u]\bigg), \tag{3.47}$$

where the continuous-normalization factor ϵ_c ensures the whole scope unitary integration. The discrete version (Eq. 3.48) is the sum of density values stored in the indexing list as

$$P_\Gamma([a_0, a_1]) = \frac{1}{\epsilon_c} \int_{a_0}^{a_1} D_\Gamma(a)\delta a, \tag{3.48}$$

where the summation limits are estimated by indexing searches (as presented in Sect. 3.4.5 in Eq. 3.41). The discrete normalization factor ϵ_d differs from its continuous counterpart ϵ_c only by magnitude but plays the same role, ensuring the (whole scope) unitary integration.

Local PDF Search Queries

The coupling between the world model representation and the visual-features requires the management of the uncertainty of the visual perception. In order to achieve this, two elements are required: First, a ground truth uncertainty model of the visual perception is needed. The role of this model is to estimate the spatial uncertainty distribution of the visual-features, namely the continuous confidence profiles of the extracted visual-features. Second, the matching of visual-features with shape-primitives must integrate the visual uncertainty model to simultaneously and efficiently compute multiple correspondences. The mechanism extracting these coupling hypotheses requires the selection and ranking of shape-primitives according to the perceptual uncertainty. A process selecting and locally ranking the shape-primitives is called *local PDF search query* $\mho : \ell_\gamma \mapsto (u_{min} \leq u \leq u_{max}) \in \mathbb{N}^+$ and is parametrized by a profile selection criterion ℓ_γ. This criterion is a 3-tuple (similar to Eq. 3.40)

$$\ell_\gamma(\underbrace{a \in \mathbb{R}^+}_{\text{perceptual cue}}, \underbrace{\gamma(a, \varsigma \in \mathbb{R}^+) \mapsto (0 \leq \delta_a \leq 1) \in \mathbb{R},}_{\text{uncertainty profile PDF at location } a} \underbrace{\delta_0 \in \mathbb{R}^+}_{\text{minimal density}}), \tag{3.49}$$

where the perceptual cue a is used to find the shape-primitives within the attribute space. Next, the profile function γ is obtained from the uncertainty model, namely the result of a particular query at the perceptual location a (see Sect. 6.7.2). Notice that the uncertainty profile function varies from one depth to another, this is the reason why the uncertainty spreading and perceptual correction must be locally estimated (see Sect. 6.7.2). Finally, the minimal probability density which is still within the aimed confidence level is denoted as δ_0. This formalization considers a general profile function γ instantiated at each particular combination of attribute and shape-primitive. An example of local PDF search query in the context of edges sorted by length using a Gaussian profile can be expressed as

$$\gamma(\overbrace{a}^{\text{perceptual cue}}, \underbrace{\varsigma(a)}_{\text{model}}) = \mathbf{exp}\left[-\frac{[\overbrace{\varsigma}^{\text{deviation}} - \overbrace{f_\mu(a)}^{\text{perceptual correction}}]^2}{2\underbrace{f_\sigma(a)^2}_{\text{uncertainty spread}}}\right], \tag{3.50}$$

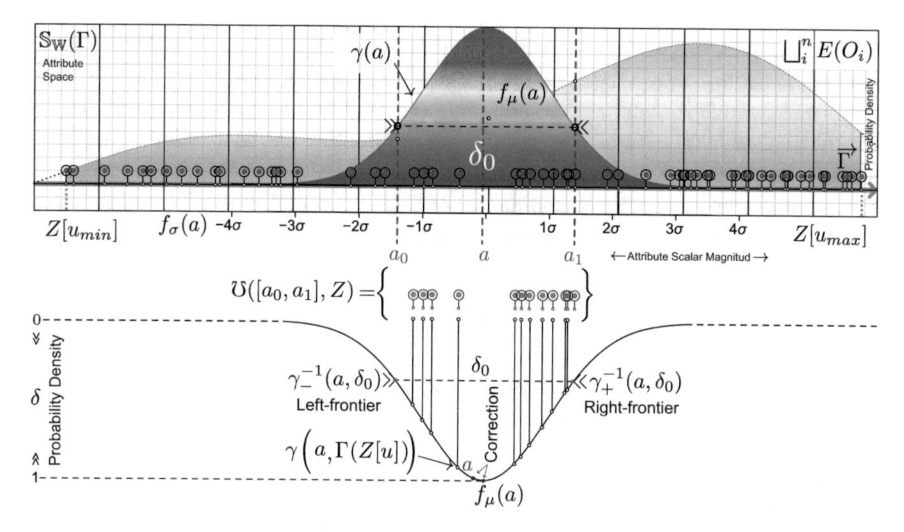

Fig. 3.20 A local probabilistic search query $\mho$ is defined by the profile selection criteria ℓ_γ composed by the perceptual cue a, uncertainty profile function γ at the location a and minimal density δ_0. The selection of the elements in the sorted list Z include their associated density values according to their evaluation in γ. Notice that the search interval is delimited by determining the frontier values a_0 and a_1 (from Eqs. 3.51 and 3.52 respectively). The γ evaluation of the local probabilistic search is shown at the lower part

$$a_0 = \gamma_-^{-1}(a, \delta_0) = f_\mu(a) \underbrace{-\sqrt{-2 f_\sigma(a)^2 \ln(\delta_0)}}_{\text{left frontier displacement}}, \tag{3.51}$$

$$a_1 = \gamma_+^{-1}(a, \delta_0) = f_\mu(a) \underbrace{+\sqrt{-2 f_\sigma(a)^2 \ln(\delta_0)}}_{\text{right frontier displacement}}, \tag{3.52}$$

$$\mho\left[\ell_\gamma\left(a, \gamma, d_0\right), Z\right] = \mho([a_0, a_1], Z), \tag{3.53}$$

where the systematic error curve (also called *perceptual correction*) $f_\mu(a)$ and spread behavior (standard deviation) $f_\sigma(a)$ are provided by the uncertainty model. The lower a_0 and upper a_1 interval boundaries are determined by the local inverse profile function at the minimal density δ_0 respectively from the left $\gamma_-^{-1}(a)$ and right $\gamma_+^{-1}(a)$ sides (see Fig. 3.20).

Query Composition

The conclusive matching of visual-features with shape-primitives requires the coupling of (at least) four points in *general position* [55]. In order to achieve the associations containing these points, it is possible to formulate queries which simultaneously restrict the DoFs of the association candidates and increase the discriminant selection of visual-features. This query composition is accomplished by relational calculus. In particular, boolean operators such as conjunction or disjunction. Due to its nature, the conjunction composition is of great importance. The aim of generating such boolean composed queries is supported by the fact that conjunction composition reinforces (drastically reduces) the possible association candidates. The association hypothesis are pairs consisting of one visual-feature and one shape-primitive. This mechanism is efficiently implemented by calculating only the minimal amount of selection of elements (from Eq. 3.42). This can be formally expressed as

$$\mathcal{Q}_{\cap}\Big([a_0, a_1], [a_2, a_3]\Big) := P_{\Gamma}\Big\{\mho([a_0, a_1], \mathrm{Z})\Big\} \cap P_{\Gamma}\Big\{\mho([a_2, a_3], \mathrm{Z})\Big\}, \tag{3.54}$$

where the notation $\{\cdot\}$ represent a sublist from Z and the special conjunction operator $\cap$ constructs a set of 3-tuples of the form

$$P_c := \Big(\underbrace{u_x}_{\mho([a_0,a_1],\mathrm{Z})}, \underbrace{u_y}_{\mho([a_2,a_3],\mathrm{Z})}, \underbrace{P_{\Gamma}(\mathrm{Z}[u_x]) \cdot P_{\Gamma}(\mathrm{Z}[u_y])}_{\text{conjunctive likelihood}} \Big) \in \mathcal{Q}_{\cap}. \tag{3.55}$$

The conjunctive composed query $\mathcal{Q}_{\cap}$ contains edge pairs satisfying length criteria. The 3-tuple also contains the combined likelihood. The query composition can use the global and local PDF search queries to achieve complex likelihood searches. Moreover, the composition of queries can incorporate additional operators. These operators can use other attributes of the shape-primitives. For instance, the conditional connectivity of boundary primitives (from Eqs. 3.33–3.38). In Fig. 3.21, the selection of two edges $e_{\alpha\beta}$ and $e_{\alpha\chi}$ connected by a common vertex v_{α} is shown. This creates selecting procedures based on probabilistic and tuple relational calculus. This exceeds the capabilities found in the database structured query language SQL (see Sect. 3.5). The composed queries exploitation is closely related to the matching approach.

3.4.6 *Conversion from CAD Models to $\mathbb{W}$-Representation*

In the last two decades, there has been a considerable development of CAD methods towards design and production of all classes of objects, houses, buildings and even the urbanization process. This trend in CAD systems yields to the development of sophisticated tools and assorted formats for interchange of CAD models. Today it is possible to exchange digital models without loss of information. This is possible due to the open standardization of the interchange file formats, for example *COLLADA* [78]. Although there is a wide variety of interchange formats for different CAGM purposes, the main targets of these computational representations are the interactive computer aided design and visualization (CAD) and the computer aided manufacturing (CAM). These two application domains are related to sensor-based recognition applications. The information structures in these domains are not

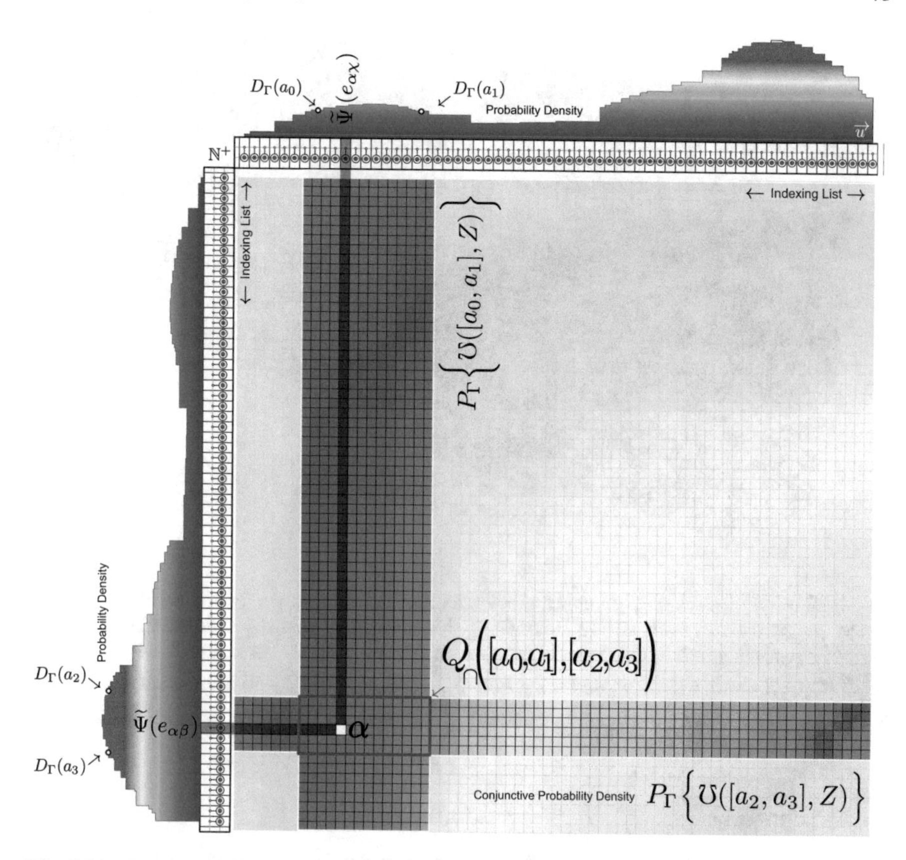

Fig. 3.21 Example of a conjunctive composed global probabilistic search query. On the top, each element in the indexing list Z stores three scalar values: (i) A reference to the shape-primitive instance. (ii) An attribute value from the indexing predicate $\Gamma(e_{\alpha\beta})$. (iii) The associate density obtained by kernel density estimation $D_\Gamma(\Gamma(e_{\alpha\beta}))$. The range determined by the indexing searches (Eq. 3.41) with arguments a_0 and a_1 is highlighted. The integration of the density in the interval approximates the ϵ_d normalized probability $P_\Gamma([a_0, a_1])$. On the left, the indexing list Z is placed orthogonal to itself in order to display the conjunctive Cartesian product in the composed query $Q_\cap$. For visualization purposes, the computation of all combinations is displayed. However, during a query only the elements within the red rectangle are computed. Finally, the composition by means of conditional queries is shown by the selection of two edges $e_{\alpha\beta}$ and $e_{\alpha\chi}$ connected by a common vertex v_α. This effective mechanism combines the probabilistic analysis with the graph structure of the world model

(at least directly) adequate for use in sensor-based systems such as visual perception for robots. The solution to this inconvenience is to convert the information provided by the interchange files in the proposed world model representation $\mathbb{W}$. This process has to be carefully done from the initial exportation of the CAD or CAM models to the importing into the world model representation and its serialization in a convenient format for sensor-based recognition. In the following, these stages are presented and discussed.

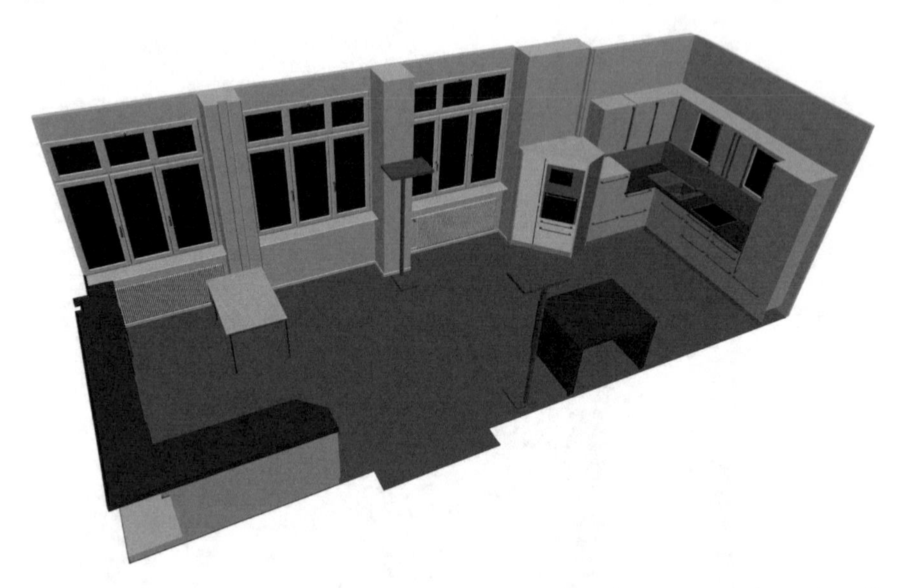

Fig. 3.22 CAD model of a complex robot kitchen. The first of the two experimental application setups used to evaluate the methods of this work. The setup is a functional full-fledge kitchen environment. This CAD model consists of $|\mathbb{W}| = 1,033$ object models with a total of $\sum_i^{|\mathbb{W}|} |C(O_i)| = 10,685$ composed faces from $\sum_i^{|\mathbb{W}|} |F(O_i)| = 58,480$ triangular faces defined by $\sum_i^{|\mathbb{W}|} |E(O_i)| = 89,767$ edges connecting $\sum_i^{|\mathbb{W}|} |V(O_i)| = 32,262$ vertices (see Table 3.3)

CAD-Exporting

There are various aspects affecting the accuracy, complexity and aggregation (see Sect. 3.1) of the object models in a CAD interchange file, namely (i) model consistency, (ii) surface subdivision and (iii) semantic and functional object segmentation. Since the two CAD environmental models (see Figs. 3.22 and 3.23) used in the experimental evaluation of this work were manually created, these three interchange aspects were carefully controlled at the generation of the models and verified by the algorithms presented in Sect. 3.4.6.

The developed methods and their implementation into a portable software system must be capable of using existing CAD models from other sources. In order to anticipate this issue, a software architecture has been proposed and implemented to enable the integration of additional CAD formats by implementing only a minimal file-loader module which maps particular ordering and specific features of additional formats. This is required for other file formats rather than `IV` *Open Inventor* or `WRL` *Virtual Reality Modeling Language*. A workaround is to convert other non-supported formats[26] to `IV` or `WRL` formats by means of open source tools.[27] Once

[26] `PLY`, `STL`, `OBJ`, `3DS`, `COLLADA`, `PTX`, `V3D`, `PTS`, `APTS`, `XYZ`, `GTS`, `ASC`, `X3D`, `X3DV` and `ALN`.

[27] http://meshlab.sourceforge.net.

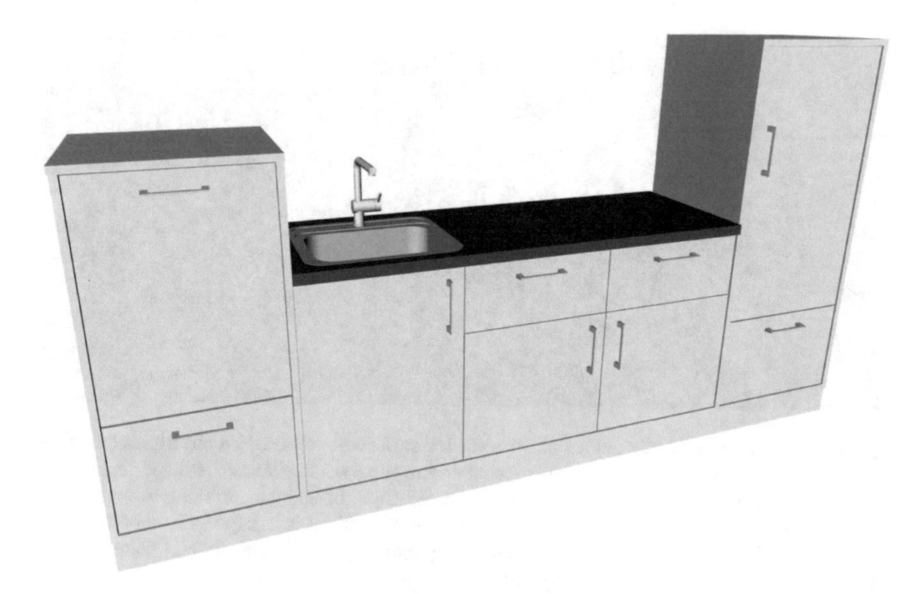

Fig. 3.23 The CAD model of a simple minimal mobile kitchen. This experimental setup is a minimal transportable kitchen with an operational sink, dishwasher and refrigerator. The CAD model consists of $|\mathbb{W}| = 50$ object models with a total of $\sum_i^{|\mathbb{W}|} |C(O_i)| = 1,582$ composed faces from $\sum_i^{|\mathbb{W}|} |F(O_i)| = 6,438$ triangular faces defined by $\sum_i^{|\mathbb{W}|} |E(O_i)| = 9,657$ edges connecting $\sum_i^{|\mathbb{W}|} |V(O_i)| = 3,325$ vertices (see also Table 3.3)

the formatting is done, the analysis and assertion of model consistency, surfaces subdivision and object segmentation is managed as follows.

Model Consistency

The most common use of CAD models is interactive design, visualization and extraction of scalar properties such as volume, areas and lengths. There might be situations in which the models have two overlapping faces or some edges have only one incidence surface instead of two (called *dangling* edges in [47, 79]). For example, window glasses (in the Fig. 3.22). These issues are not critical for the aim of visualization. However, when converting a CAD model from an interchange file exposing these issues, the resulting world model representation $\mathbb{W}$ will be damaged because the formal definitions of $\mathbb{W}$ are not valid, more specifically the properties defining the homogeneous 2D topologic polyhedron discussed in Sect. 3.4.4 are not fulfilled.

This causes inconsistencies during querying, filtering or even worse, it yields to unpredictability states during pose estimation. The detection of such irregularities is done straightforward by a set of conditional verifications based on the $\mathbb{W}$ formalism (in Eq. 3.9 up to Eq. 3.34). However, the automatic correction of these issues is not a trivial problem. In the two CAD models used for the evaluation of the proposed meth-

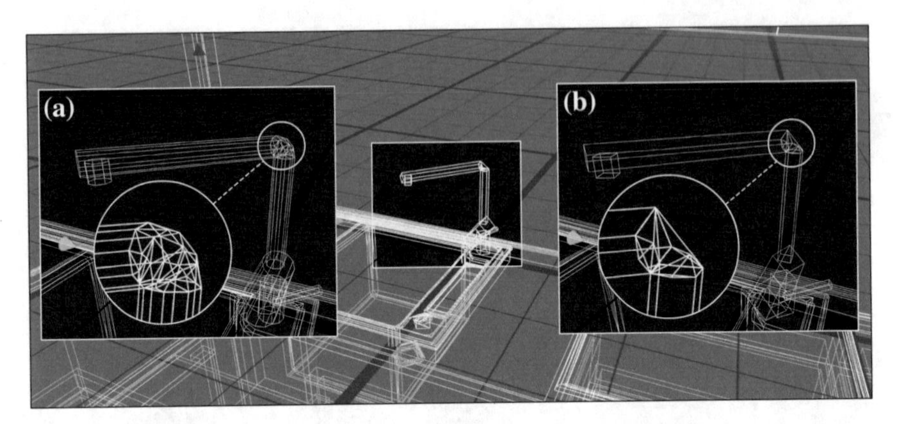

Fig. 3.24 The effects of two angles of aperture for surface approximation are illustrated at the water faucet. This element is drastically affected during subdivision because its shape creates very deviating polygonal approximations depending on the angle of aperture. **a** The narrow angle of aperture (fine granularity) generates accurate shape approximations with large amount of polygons. The narrow angle of aperture generates non-perceptible edges and faces. **b** The wide angle of aperture reduces the amount of shape-primitives

ods, the detection of these irregularities is preformed by software and the corrections were done manually using available 3D-modeling tools.[28]

Surface Subdivision

Another important aspect while exporting CAD models for their conversion or inclusion into the world model representation $\mathbb{W}$ is the surface subdivision. This substantial aspect is controlled by the angular subdivision threshold. This is the angle which corresponds to the minimal aperture (see Eq. 3.19) for a curve to be divided into two different segments. For example, the Fig. 3.24 shows two subdivisions of the same object model with different angle of aperture. The selection of this angle depends on the target application. This means, a trade-off between resolution and saliency of the subdivision faces. For instance, when the visual localization of an element is intended, the angle is rather wide, whereas the angle must be kept narrow for applications involving manipulation (see the subdivision used in models for grasp-planning in [80]) in order to properly plan stable contact points. The angle of aperture is of special interest when using visual-features which are sensitive to curvature, for instances edges. In this work, the angle of aperture was selected at both limits (see Fig. 3.24) and exported in two versions, a low-resolution (wide angle of aperture) and high-resolution (narrow angle of aperture) CAD interchange files (see details in Table 3.3). This was done with the purpose of quantitatively evaluating this aspect while qualitatively analyzing the effects of the surface subdivision (see Sect. 3.5).

---

[28]CAD models were modeled with *ProENGINEER* (Wildfire 4.0®) and subsequently exported to the *Open Inventor* `IV` format (see Sect. 3.4.6).

Object Segmentation

The segmentation of elements composing the environment into encapsulated objects modeled and stored in CAD representations requires criteria which consider spatial relations, functional dependencies and semantic associations of the elements. The formulation of these criteria based on rules and heuristics can support the autonomous acquisition of complex models. This approach has yielded and some promising results (see [64, 81]) which support the use of CAD representations acquired by active-sensors. In this representation, the spatial object segmentation is done with the boundaries of the physical composition of walls, doors, cupboards and shelves (see Fig. 3.25). The functional aggregation of the elements has been realized manually. For example, the location of the rotation axis doors or the translation vector of drawers were added during the model creation. The semantic information in the world model representation are the names of the object models with rotational or prismatic joints such as electrical appliances, doors and drawers.

$\mathbb{W}$-Importing and Storing

Once the CAD interchange file has been properly generated for sensor-based recognition, import the models into the world model representation has to consider various aspects of the CAD files (see Fig. 3.26). The redundancy of the elements and additional information for visualization (called shape hints) for smooth surfaces, materials and units must be consider and filtered. The following sections discusses these aspects leading to important considerations for the redundancy removal and extraction of composed faces. Finally, the serialization and storage of the world model representation are discussed and illustrated.

Vertex Redundancy

A notable characteristic of the serialization of CAD files is the redundancy in terms of vertex repetition. This occurs because the list of shape primitives (usually triangles, strips, fans or convex polygons such as quads) enumerate the vertex coordinates in a sequential order. For example, in the object model of Fig. 3.13, the coordinates of the vertex v_4 appear five times as follows: $f_{1,4,2}$, $f_{2,4,3}$, $f_{1,8,4}$, $f_{3,4,7}$ and $f_{4,8,7}$ (see example with seven occurrences in Fig. 3.27).

This vertex redundancy occurs because the explicit and redundant storage of the coordinates (in the device memory either CPU or GPU) notably improves the rendering performance (shading, occlusion culling, ray tracing) and other operations necessary for visualization. In the proposed representation, each vertex is unique and its linkage has to be stored for the graph incidence (as introduced in Sect. 3.4.4). The generation of a table with unique vertex identifiers and coordinates helps to fuse close located vertices (within a small threshold radius) and identify multiple references to the same vertex. This determines whether a vertex has been listed before. This simple operation implies a comparison of vertices while loading the object models from the interchange file. This noise tolerant comparison requires many floating point operations (calculating the l^2-norm) and the vertices have many occurrences. Thus, the process needs considerable time for computation (see details in Table 3.3).

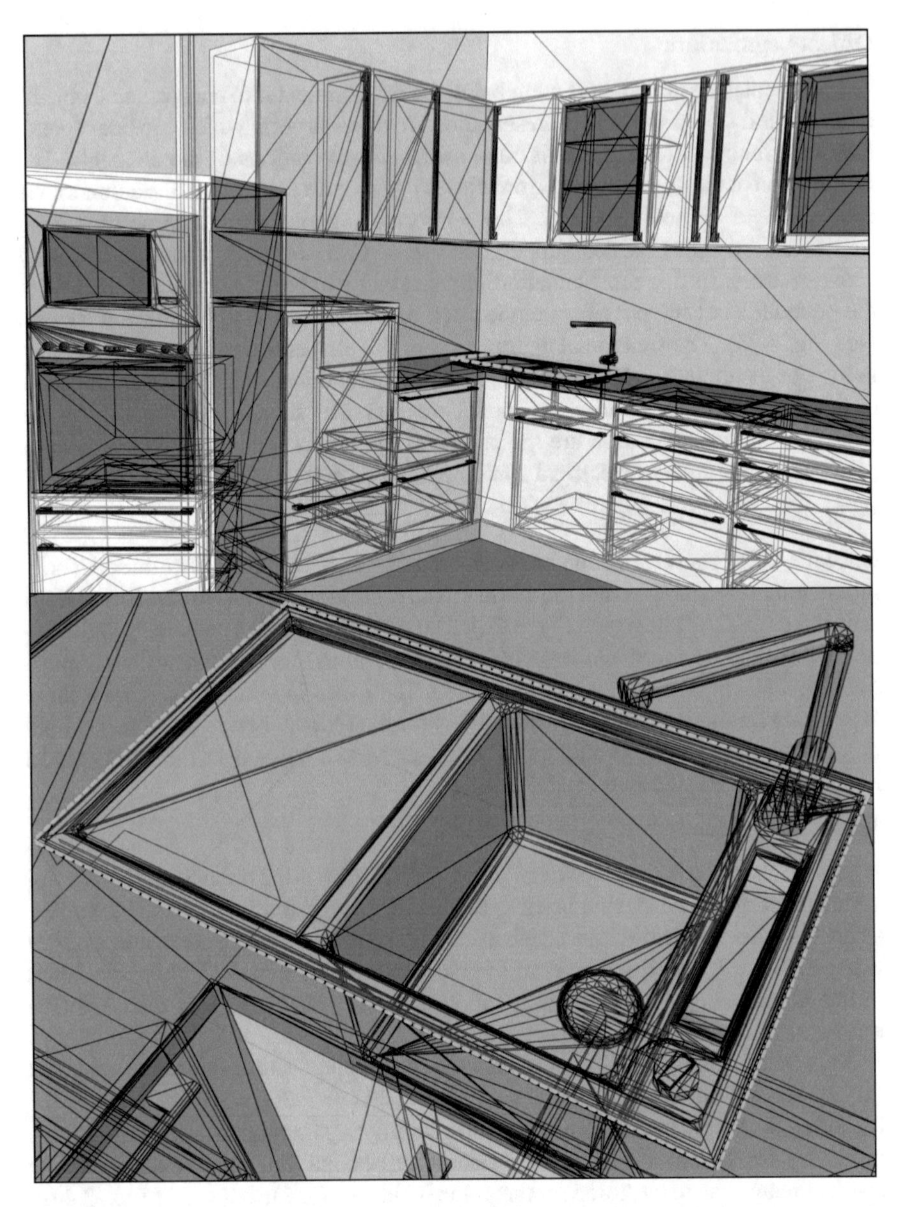

Fig. 3.25 Object separation by geometric, functional and semantic criteria. The robot kitchen environment shows the composition of environmental elements. The segmented and labeled sink shows a composition ideal for illustrating redundancy reduction, surfaces division and visual saliency detection

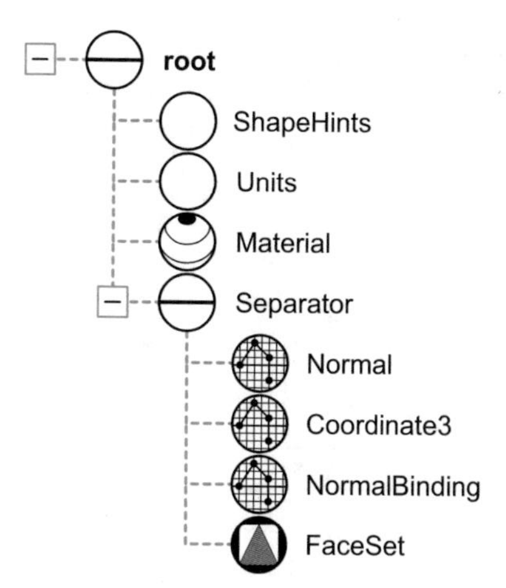

Fig. 3.26 The *Open Inventor* is a scene graph and CAD format. This subgraph is a visualization of the first nodes of the model in Fig. 3.22

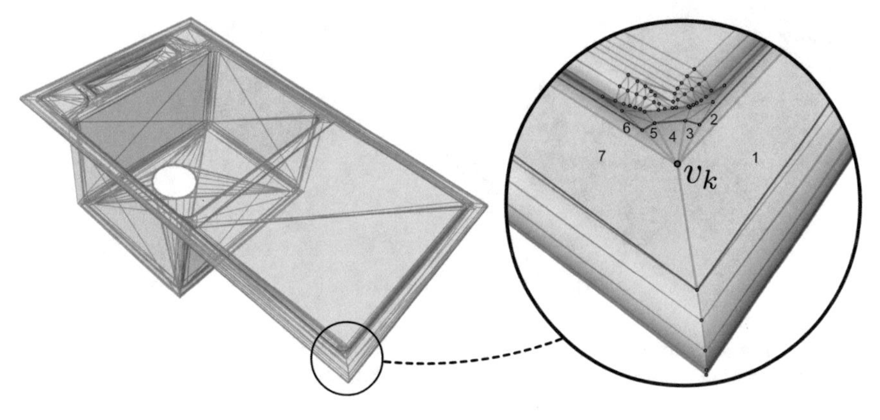

Fig. 3.27 An example of a vertex v_k with seven occurrences in the CAD model. This repetitive occurrences are removed for graph indexing (see Sect. 3.4.4). The green lines show the incident edges to v_k (see Eq. 3.33) and the red lines denote edges associated by face incidence (see Eq. 3.35)

Edges Redundancy

In Sect. 3.4.4 an important fact has been discussed: The visibility of virtual edges (Eq. 3.31). This equation states, there are edges whose only function is to act as auxiliary elements for the surface subdivision. This means, that these edges are not perceptible. Consequently, this attribute can be used to detect which edges are considered for the indexing (as presented in Sect. 3.4.5). The Fig. 3.28 illustrates these observations. Disregarding the virtual edges is actually the first step towards a non-redundant and visually representative model. The next step is to avoid edges which are not robustly perceptible due to their short length or non-salient aperture. Thus, length filtering is performed by selecting a lower and upper length thresholds

Table 3.3 Transformation effects of the angle of aperture for surface subdivision. The two setups were exported with the maximal and minimal angles of aperture producing different representations. The connectivity increment in terms of edges in the high resolution robot kitchen is more than six times the size of the low resolution version. The ratio between the total amount of composed faces $\sum_i^{|\mathbb{W}|} |C(O_i)|$ (denoted as C-Faces) and the triangular faces $\sum_i^{|\mathbb{W}|} |F(O_i)|$ (expressed as T-Faces) is higher for the low resolution versions. This occurs because the narrow angle of aperture forbids the fusion of small partial surfaces into larger composed faces

$\mathbb{W}$-Model	Aperture	Models	Vertices	Edges	T-Faces	C-Faces
Robot-	$\pi/2$	1033	33411	14408	23758	5982
Kitchen	$\pi/7$	1033	36262	89767	58480	10685
Fig. 3.22	Ratio	1.00	1.09	6.23	2.46	1.79
Mobile-	$\pi/2$	50	1875	5307	3538	1269
Kitchen	$\pi/7$	50	3325	9657	6438	1582
Fig. 3.23	Ratios	1.00	1.77	1.82	1.82	1.25

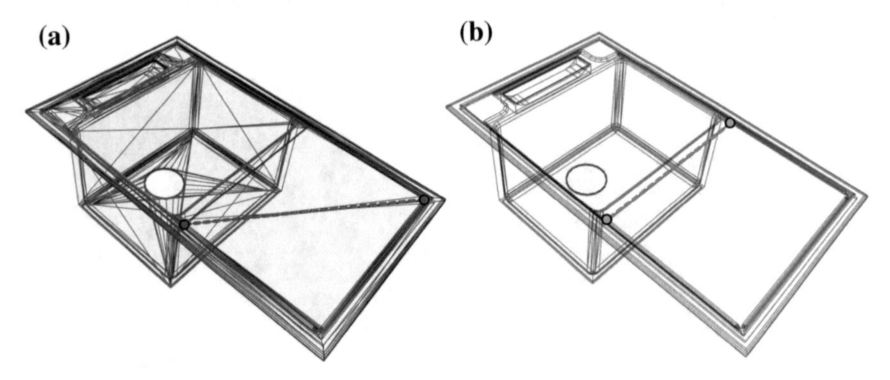

Fig. 3.28 The detection and removal of object model redundancies. **a** The virtual edges $\breve{e}_{\tau\upsilon}$ are removed for indexing and visualization. **b** This stage notably reduces the amount of edges for visual matching

(see Sect. 3.5) based on the calibrated 3D field of view of the humanoid robot (see Fig. 3.29).

Face-Vertex Oriented Incidences

When defining smooth surfaces, the representation in CAD interchange formats (such as in Fig. 3.26) provides normal vectors associated to vertices in the definition of each face (see Fig. 3.12). For example, in a triangular face $f_{\alpha\beta\chi}$ used to approximate a smooth surface, there are three different normal vectors. Each of these normal vectors corresponds to a vertex $\widetilde{N}(\alpha, f_{\alpha\beta\chi})$, $\widetilde{N}(\beta, f_{\alpha\beta\chi})$ and $\widetilde{N}(\chi, f_{\alpha\beta\chi}) \in \mathbb{R}^3$. This information is fundamental for various tasks. In particular for visual perception, this orientation vector determines the angle of aperture of the edge. Thus, the maximal angle of aperture of an edge determines whether or not the edge is considered for indexing (see Fig. 3.29).

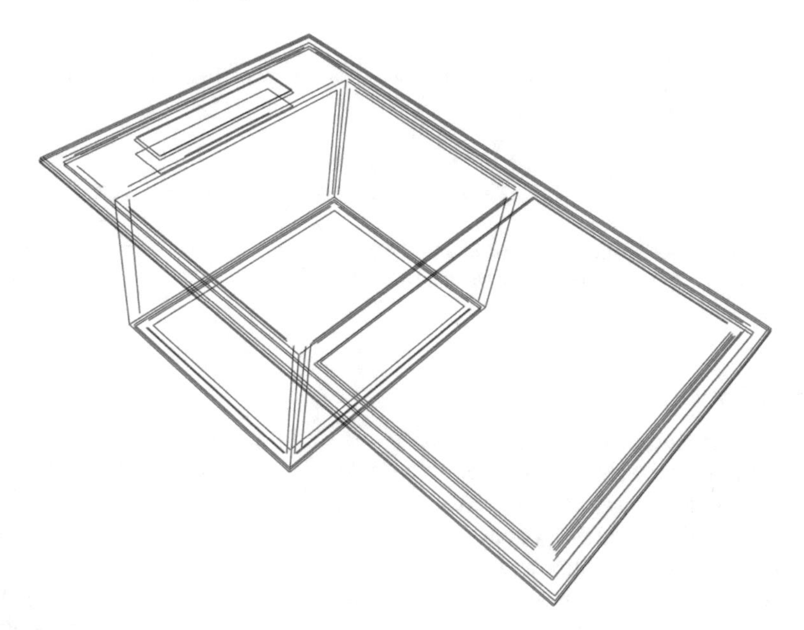

Fig. 3.29 Removal of visually indistinguishable edges (based on length and angle of aperture) filters the visually salient edges

Fig. 3.30 The composed faces $C(O_i)$ of an object model are classified by convexity. This example shows selected surfaces by area and convexity

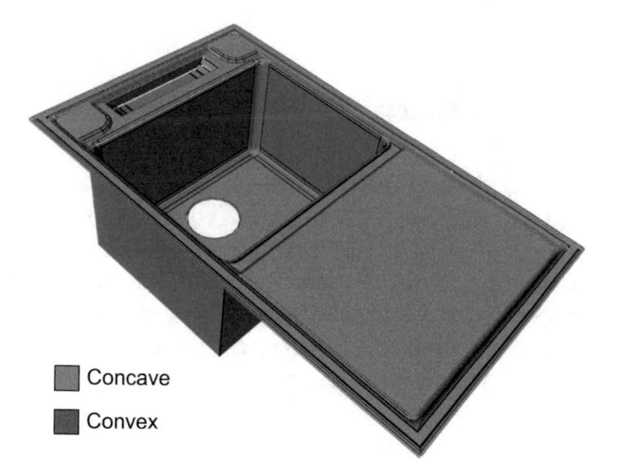

Extraction of Composed Faces

The detection of virtual edges supports the extraction of composed faces S_h (Eq. 3.24) by incremental expansion of triangular faces connected by (at least) one virtual edge. This process is efficiently done by the sequential analysis of the face set $F(O_i)$, namely verifying the faces references $K(f_{\alpha\beta\chi})$ (Eq. 3.27), common orientation (Eq. 3.25) and edge connectivity (Eq. 3.26). The resulting composed faces are analyzed to determine additional properties. For instance, in the Fig. 3.30 the surface convexity is shown.

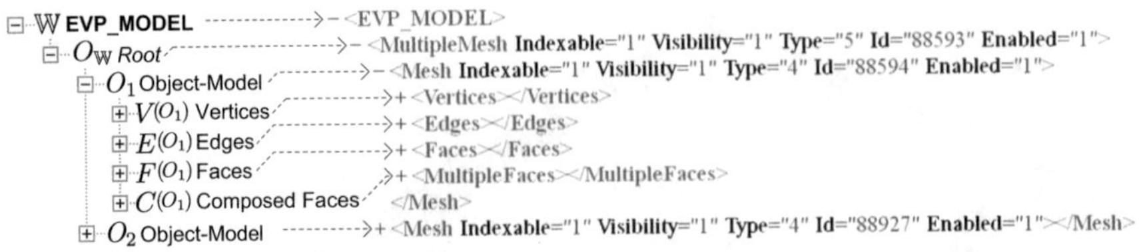

⊟ $\mathbb{W}$ **EVP_MODEL** ------> – <EVP_MODEL>
⊟ $O_{\mathbb{W}}$ *Root* ------> – <MultipleMesh **Indexable**="1" **Visibility**="1" **Type**="5" **Id**="88593" **Enabled**="1">
⊟ O_1 Object-Model ------> – <Mesh **Indexable**="1" **Visibility**="1" **Type**="4" **Id**="88594" **Enabled**="1">
⊞ $V(O_1)$ Vertices ------> + <Vertices></Vertices>
⊞ $E(O_1)$ Edges ------> + <Edges></Edges>
⊞ $F(O_1)$ Faces ------> + <Faces></Faces>
⊞ $C(O_1)$ Composed Faces ------> + <MultipleFaces></MultipleFaces>
</Mesh>
⊞ O_2 Object-Model ------> + <Mesh **Indexable**="1" **Visibility**="1" **Type**="4" **Id**="88927" **Enabled**="1"></Mesh>

Fig. 3.31 XML serialization of the world model representation. The arrows show the correspondence between mathematical concepts and implementations. This XML shows two object models from Fig. 3.23

$\mathbb{W}$-Serialization and Storage

After a CAD model has been converted into the world model representation $\mathbb{W}$, it is possible to serialize $\mathbb{W}$ into an extensible markup language. The *XML* encoding is used due to its extensibility, simplicity, generality and usability (see Fig. 3.31). The serialization allows the storage and transmission[29] of the representation avoiding the re-computations of detection and removal of redundancies, composed face extraction and indexing. The XML serialized world model representation is a direct mapping of the indices and properties of the shape-primitives. Thus, the time required to load the XML file into the memory of the visual perception system is only a fraction of the time required to re-import the model(s) from the CAD interchange file(s).

3.5 Evaluation

The evaluation of the world model representation $\mathbb{W}$ considers the setups presented (in Figs. 3.22 and 3.23). Each model was exported with two different subdivision (at both extrema $\frac{1}{2}\pi$ and $\frac{1}{7}\pi$) angle of apertures and summarized (in Table 3.3). In order to simultaneously evaluate both CAD models and the proposed world model representation, a characterization of the shape-primitives is presented (in Sect. 3.5.1). Afterwards, the indexing and analysis of the PDFs of the edges and composed faces are presented (in Sect. 3.5.2).

3.5.1 Characterization of Shape-Primitives

Vertex Characterization

Each vertex of an object model has a connectivity degree called valence $\nu(v_k) \mapsto \mathbb{N}^+$ (Eq. 3.15). This property enables the search according structural characteristics. For instance, the vertex at the corner of a box v_c should have valence $\nu(v_c) = 3$. However,

[29] Diverse binary serializations can be selected to efficiently transfer the representation.

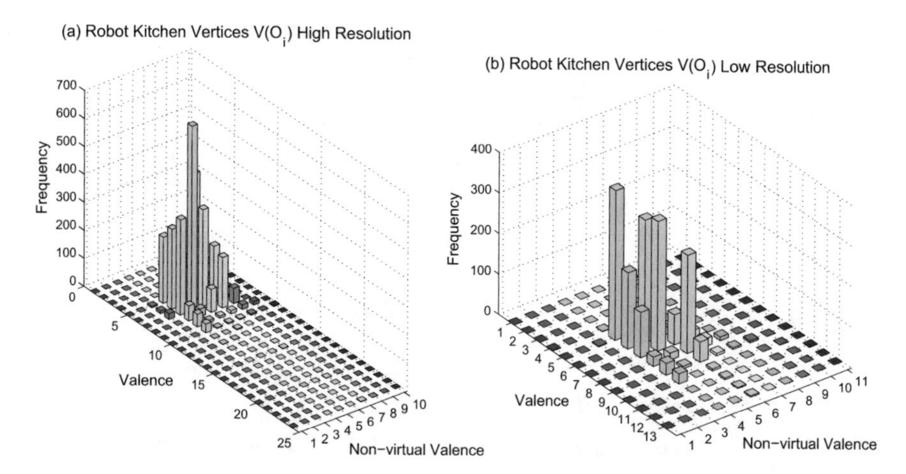

Fig. 3.32 The vertices characterization in terms of valence $\nu(v_k)$ and non-virtual valence $\hat{\nu}(v_k)$ from the robot kitchen setup in Fig. 3.22. **a** The plot shows the frequency and relation between the direct connectivity of the vertices (Eq. 3.15) and their non-virtual valence. **b** The concept is illustrated for the vertices obtained from the model with narrow angle of aperture (the high resolution version). Similar pattern arises when analyzing the mobile kitchen setup. However, since the amount of vertices is significantly smaller, the visualization is less representative and therefore omitted

due to subdivision effects, some vertices have a higher degree. The extraction of these structures requires more elaborated filtering routines. The key to unveil these structures is to omit the incidence of virtual edges $\breve{e}_{\tau v}$ (Eq. 3.31) while computing the valence of a vertex, namely the non-virtual valance $\hat{\nu}(v_k)$ (Eq. 3.23). Figure 3.32 illustrates the relation of both valences.

Edge Characterization

The two important characteristic properties of an edge $e_{\alpha\beta}$ are its angle of aperture ($\omega(e_{\alpha\beta})$ Eq. 3.20) and its length $\Gamma(e_{\alpha\beta})$ (Eq. 3.12). Based on these properties, it is possible to determine whether an edge can be perceived as salient visual-feature for recognition.

This off-line analysis enables the selection of edges which are more likely to be found in images. In Figs. 3.33 and 3.34, the detailed presentation and application of this concept is clearly presented with a comparison of the resolution effects. In each of these plots, there are two remarkable clusters. The most populated is distributed along the null angle of aperture $\omega(e_{\alpha\beta}) = 0$ and the (less salient) one along the perpendicular angle of aperture $\omega(e_{\alpha\beta}) = \frac{\Pi}{2}$. Their existence is evident when realizing that many edges are dividing either the coplanar surfaces or there are many edges arising at the intersection of perpendicular planes located at $\omega(e_{\alpha\beta}) = \frac{\Pi}{2}$, for instance the doors or drawers (see Fig. 3.23). This characterization shows an important fact. Only a small percentage of the edges (less than 11.5%) are visual significant for recognition. This amount is even smaller when considering fixed occlusions, namely the edges which are occluded from the reachable space.

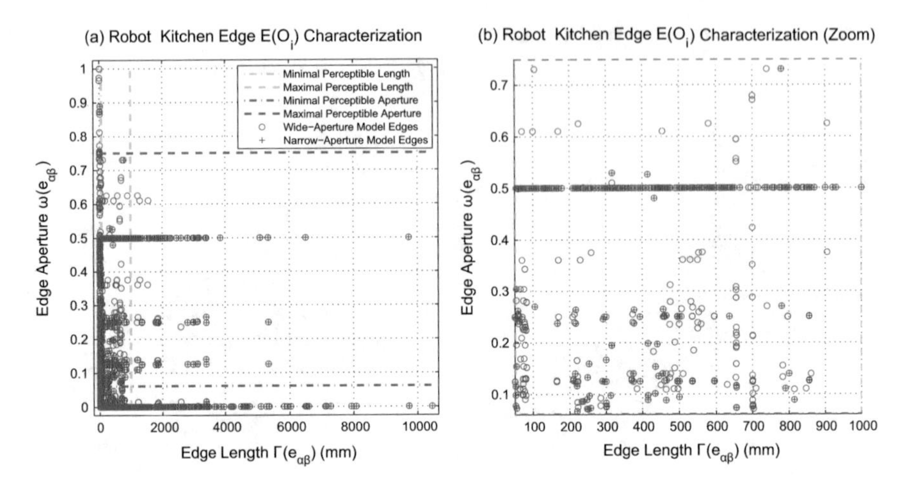

Fig. 3.33 Edge characterization of the robot kitchen Fig. 3.22. **a** Model edges are shown by two different subdivisions. Red circles illustrate edges from low resolution model. Blue crosses represent edges using high resolution model. The cyan vertical dotted lines (located at 50 and 1000 mm) delimit the visually detectable edge length. The horizontal magenta lines (placed at 0.06125π and 0.75π) delineate the discernible edge aperture. **b** The zoom of the discernible region. Only $4184 \approx 11.538\%$ of the edges are discernible in the low resolution model and $7124 \approx 7.936\%$ in the high resolution model

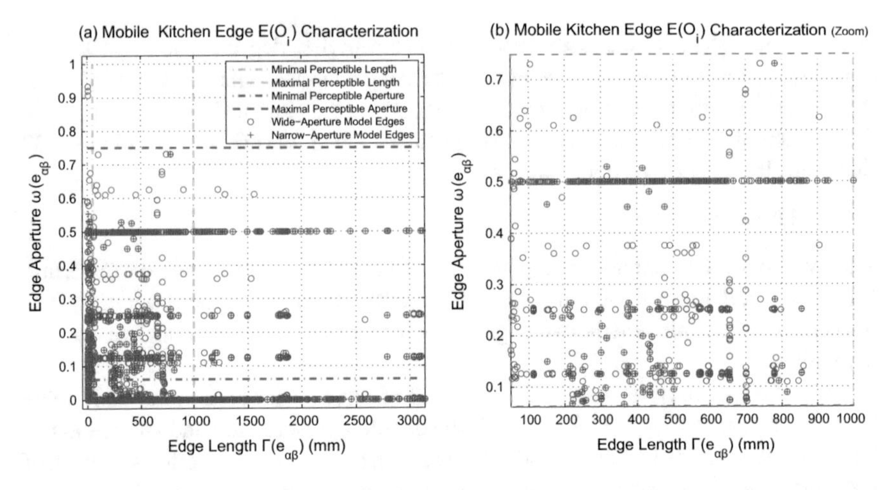

Fig. 3.34 **a** Edge characterization of the mobile kitchen Fig. 3.23. **b** Only $1034 \approx 19.483\%$ of the edges in the low resolution model and barely $1094 \approx 11.328\%$ edges in the high resolution model are discernible

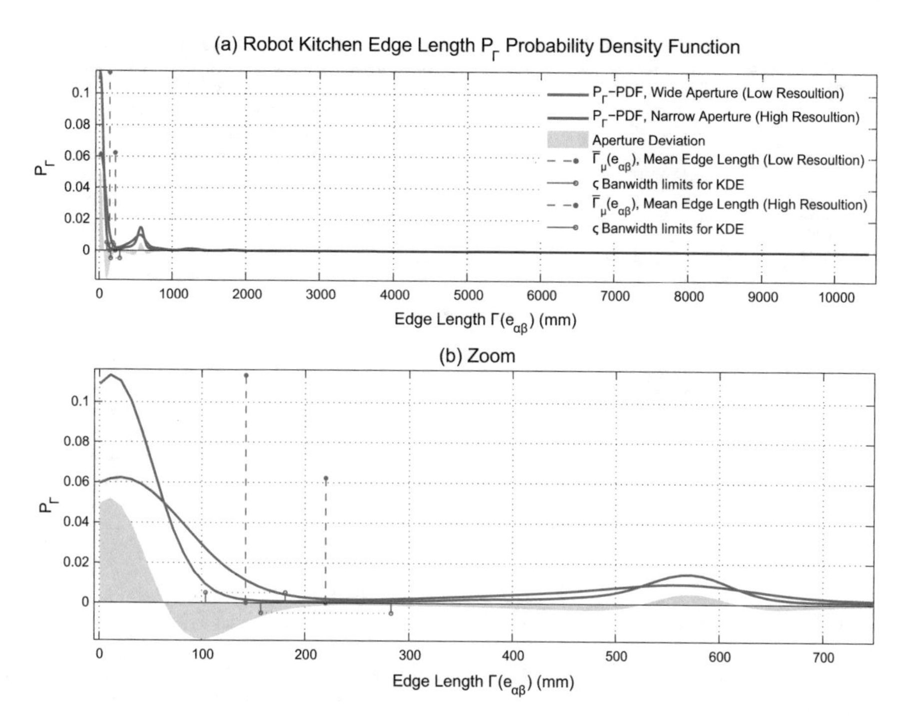

Fig. 3.35 **a** The edge length distribution (estimated by KDE, Eq. 3.43) of the robot kitchen (in Fig. 3.22) is non-parametric and multimodal. The region with minimal length has maximal density. This occurs because the small triangular faces (used to approximate curved surfaces) are dominant in terms of frequency. **b** The region zoom contains edges with length shorter than 1,000 mm. This zoom enables a better visualization of the mean edge length $\Gamma_{\mu}(e_{\alpha\beta})$ for both resolution models. The bandwidth (Eq. 3.45 and Eq. 3.46) is visualized for both resolutions

3.5.2 *Indexing and Probabilistic Search Queries*

Edge Length: The PDF of the edge length is necessary for probabilistic search queries. In Sect. 3.4.5, the PDF enables the selective filtering of edges. The PDFs (in Figs. 3.35 and 3.36) show the results obtained using both model setups. These plots are conceptually discussed in Fig. 3.18. When plotting large amounts of samples, a regular sampling of the PDF is required. For instance, in the high resolution kitchen with 89767 edges, a discrete increment step with a resolution of 3 mm was selected (see Fig. 3.35).

Face Area: As discussed in this chapter, the visual recognition of elements in the environment can be observed by the extraction and matching of supported properties such as length and structural composition. Another alternative is the recognition of

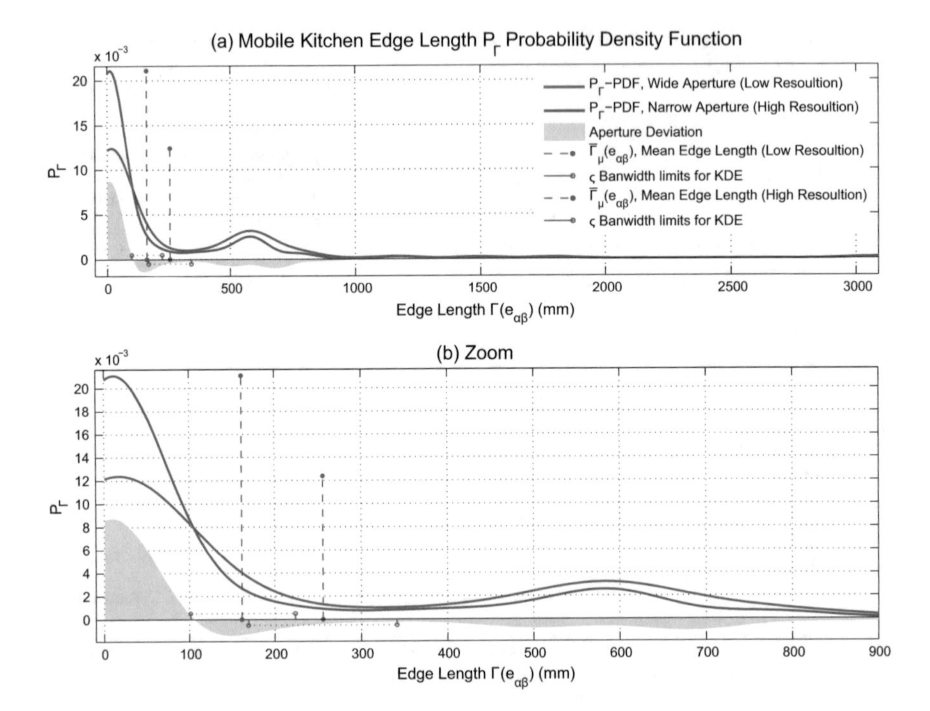

Fig. 3.36 The edge length distribution (from Eq. 3.43) of the mobile kitchen setup (in Fig. 3.23) is (in same manner as Fig. 3.35) non-parametric and multimodal. **a** The plot shows the full length distribution. **b** The plot shows the region zoom containing the distribution of the edges shorter that $\Gamma(e_{\alpha\beta}) < 900$ mm. A remarkable fact of this edge length distribution (in both setups and resolutions) are the two modes, one at the very short edge length (approximately $\Gamma(e_{\alpha\beta}) < 100$ mm) and the second located between $400\,\text{mm} < \Gamma(e_{\alpha\beta}) < 800$ mm. The second region is very significant for visual recognition because there are many elements (high frequency) which are visual salient at manipulable distances for the humanoid robot

planar polygonal faces represented by composed faces S_h. Using the area attribute for recognition requires to index and analyze the composed faces of the object models in the world model.

The Figs. 3.37 and 3.38 show the distribution of the area of the composed faces in both setups. These results simultaneously illustrate the resolution comparative (using both angles of aperture) and the interest region in the zooms. This allows to focus on the search and matching of elements which are found less often (therefore more restrictive) for a conclusive match.

Query Performance: In order to evaluate the representation in terms of time performance during the search of shape-primitives, different queries have been executed with large amount of trials (1000 per configuration). In these trials, the parameters were changed dynamically (within the scope of the particular attribute and shape-primitive) to obtain a reliable analysis of the system behavior. Based on these results,

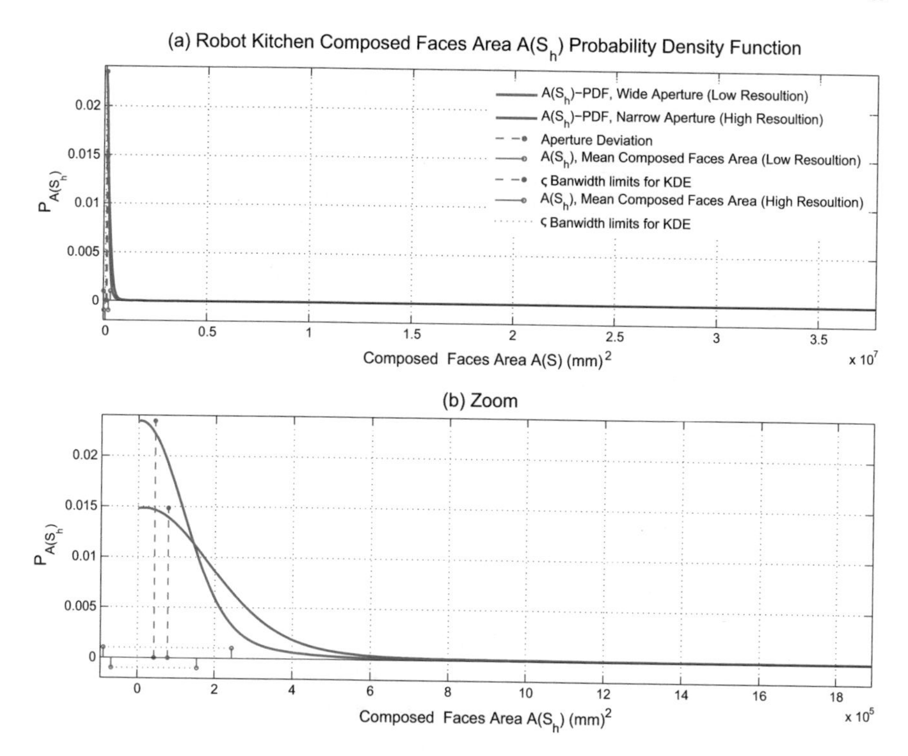

Fig. 3.37 The area distribution of the composed faces (Eq. 3.43 using the predicate function in Eq. 3.28). This particular distribution was obtained from the robot kitchen setup (see Fig. 3.22). **a** The plot shows the full area distribution. **b** The plot shows the region zoom containing the distribution of the composed faces which are smaller that $A(S_h) < 18 \times 10^5 \text{mm}^2$. The two different versions (in blue and red lines) of the same model (due to different angles of aperture) expose slightly variations. The scale is in square millimeters (mm^2). This has the effect of a wide rage in the horizontal axis

it is possible to determine which are the shortest, mean, longest and standard deviation response times resulting from the implementation of the world model representation (see Fig. 3.39).

3.5.3 Serialization and Storage

Once a CAD model has been converted into the world model representation $\mathbb{W}$, it is serialized as presented in Sect. 3.4.6. The results are shown in terms of file size, processing and loading times in Table 3.4.

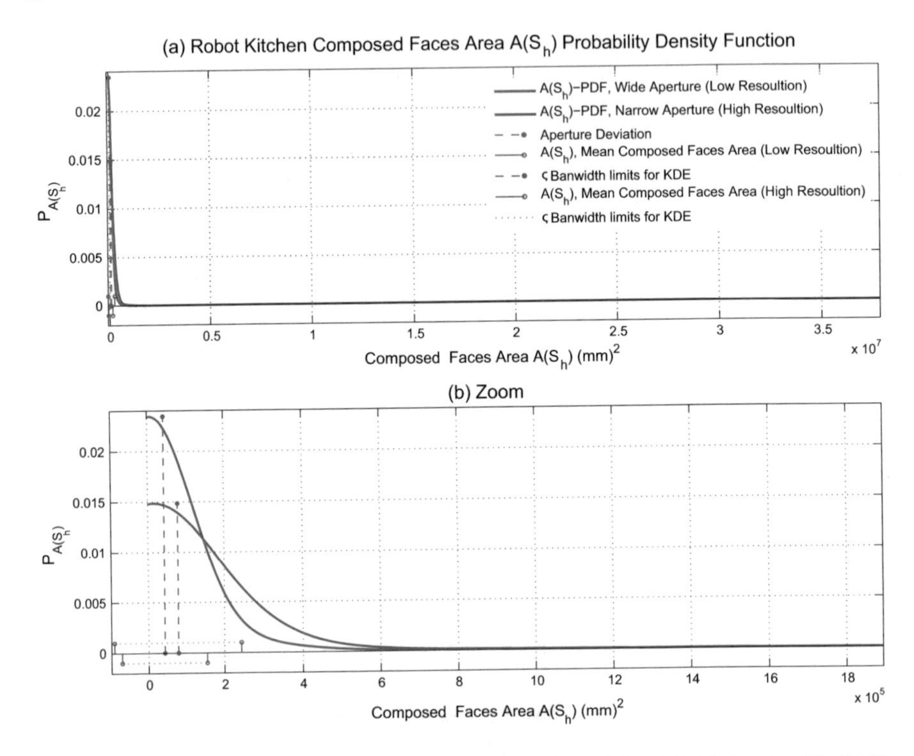

Fig. 3.38 The area distribution of the composed faces from the model mobile kitchen (see Fig. 3.23). In analogy to Fig. 3.37, the area distribution of the composed faces in the upper plot shows that most of the composed faces are in the low area region of the plot. This occurs because the surface subdivision of all rounded chamfers (all over the doors and surfaces of the model) generate small rectangles which are over representative for the PDF. This is also an important criterion to search for faces whose area is beyond this region

Table 3.4 The serialization of the world model representation $\mathbb{W}$. Results using the narrow and wide angles of aperture versions of both models in terms of storage size. The processing time required to import (from `IV` format) and generate the representation is given. Finally, the time necessary to load the serialization for recognition modules is provided

$\mathbb{W}$-Model	Aperture	CAD_{kb}	XML_{kb}	$Filter_{\text{ms}}$	$Loading_{\text{ms}}$
Robot kitchen Figure 3.22	$\pi/7$	19397	41327	3300	1950
	$\pi/2$	7889	16944	1575	1470
Mobil-kitchen Figure 3.23	$\pi/7$	1867	4284	746	740
	$\pi/2$	975	2370	192	160

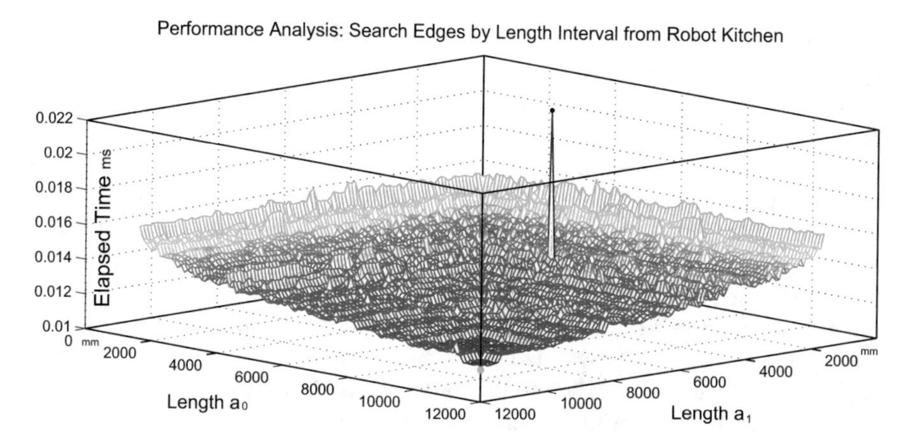

Fig. 3.39 The time performance of the search interval (Eqs. 3.41 and 3.42). The plot shows in the horizontal axes the distances a_0 and a_1 (the lower and upper boundaries) of the interval search. The vertical axis is the maximal elapsed time. Each configuration was evaluated with 1000 trials. This large amount of searches enables to determine the shortest 0.0109 ms, longest 0.0213 ms response times. The mean performance time is 0.0128 ms with an standard deviation of 34.392 μs

3.6 Discussion

In this chapter, a formal and computational world model representation for environmental visual perception for robots was introduced. The proposed representation provides coherent and proficient descriptions of physical entities found in made-for-humans environments. The representable scope of this graph-based representation is general without imposing sensor-dependence, neither requiring complex instance specific training. These guidelines ensure the wide collaborative integration of diverse sensing modalities for robots sharing portable and network-accessible world model representations.

Particularly, in order to be used in vision-based recognition, the high sensitivity and uniqueness criteria are mandatory for the proposed world model representation. These criteria were held by geometric shape-primitives capable of reflecting subtle differences with adjustable granularity by surface subdivision. The geometric shape-primitives expose these characteristics with remarkable stability and accuracy. Due to the compact and generalized nature of these geometric shape-primitives, the model-based geometric representation is efficient in terms of processing and storage. This efficiency was achieved by a topological composition. In this manner, geometric shape-primitives can be hierarchically and multimodally organized in order to arrange vast amounts of models. The sustaining of such hierarchical composition is positively reflected into high scalability, whereas the multimodal organization based on indexing schemes allows optimal and complex attribute-based search procedures.

In addition, the object-centered nature of the model-based geometric representation enables the storage of extensible data structures necessary for probabilistic perception, semantic planning and symbolic reasoning. The resulting extensibility also provides means to handle dynamic parametric transformations present in plenty of environmental objects. These hierarchically organized transformations are ideal for depicting complex aggregations in household environments such as electric appliances or manipulable furniture. The selected boundary geometric elements such as vertices, edges and polygonal faces can be conveniently used as local shape-primitives. This is the key to manage occlusion and self-occlusion in environmental visual perception. These achieved fundamental criteria are based on the analysis of four decades of research on object representations for visual detection and recognition. Section 3.3 presented a selection of researches supporting these arguments. Particularly important is the qualitative comparison in Sect. 3.3.10 which is summarized in Tables 3.2 and 3.1. These statements are supported by the evaluation of the proposed representation using two different setups in Sect. 3.5.

Although these arguments and evidence clearly support the use of a model-based representation approach, a solely geometric model-based representation cannot soundly address all visual perception task for robots. The limitations of geometric model-based representations are:

- **Circumscription**: Model-based approaches imply a closed-world which cannot be hold in all made-for-humans domains [82].
- **Changes**: The dynamic modifications of the world either *transient* or *lasting* [83] cannot be thoroughly represented by pure geometric models. This occurs when the alterations of the objects forbid their soundly and enduring representation.
- **Composition**: The limited representativeness or granularity of the models restrain the representable objects [84]. For instance, plants, food, etc. (see also [3]) cannot be soundly and efficiently represented by CAGM methods.
- **Consistency**: The inherent modeling inconsistencies between models and real instances complicate the association between visual-features and model descriptions. This can lead to negative effects.

A pure model-based approach would insufficiently address these issues. The comprehensive solution—would at least—require a combined egocentric and allocentric visual and spatial memory [85] which must include not only the models but also their spatio-temporal varying transformations (including some sort of appearance information) of all recognized and registered model instances.

Based on previously discussed facts, a model-based visual perception approach can provide bounded but solid basis required to achieve the essential environmental visual awareness for autonomous intelligent robots. This basis has to supply answers to the *what* and *where* questions (discussed in Sect. 1.1) within the assumed closed-world model. The development of the basis for comprehensive visual perception requires the answers to visual queries while reliably managing the natural conditions in real made-for-humans scenarios with non-controlled illumination, complex structures and materials. These conditions place the challenges at the robustness, precision, scalability and wide applicability of the proposed approach. In the following chapters, novel methods for visual sensing and feature extraction cope with this challenges by means of image fusion and edge graphs representations.

References

1. Pope, A.R. 1994. *Model-Based Object Recognition - A Survey of Recent Research*. University of British Columbia, Vancouver, BC, Canada, Canada, Technical Report.
2. Mundy, J. 2006. Object Recognition in the Geometric Era: A Retrospective. In *Toward Category-Level Object Recognition*, ed. J. Ponce, M. Hebert, C. Schmid, and A. Zisserman, 3–28. Lecture Notes in Computer Science. Berlin: Springer.
3. Gonzalez-Aguirre, D., J. Hoch, S. Rohl, T. Asfour, E. Bayro-Corrochano, and R. Dillmann. 2011. Towards Shape-based Visual Object Categorization for Humanoid Robots. In *IEEE International Conference on Robotics and Automation*, 5226–5232.
4. Flint, A., C. Mei, I. Reid, and D. Murray. 2010. Growing Semantically Meaningful Models for Visual SLAM. In *IEEE Conference on Computer Vision and Pattern Recognition*, 467–474.
5. Krueger, N., C. Geib, J. Piater, R. Petrick, and M. Steedman, F.W. örgötter, A. Ude, T. Asfour, D. Kraft, D.O.A. Agostini, and R. Dillmann. 2011. Object-Action Complexes: Grounded Abstractions of Sensory-motor Processes. *Robotics and Autonomous Systems* 59 (10): 740–757.
6. Thrun, S., W. Burgard, and D. Fox. 2005. *Probabilistic Robotics, Intelligent Robotics and Autonomous Agents series*. Intelligent robotics and autonomous agents: The MIT Press. ISBN 9780262201629.
7. Davison, A., I. Reid, N. Molton, and O. Stasse. 2007. MonoSLAM: Real-Time Single Camera SLAM. *IEEE Transactions on Pattern Analysis and Machine Intelligence* 29 (6): 1052–1067.
8. Roth, P.M., and M. Winter. 2008. *Survey of Appearance-based Methods for Object Recognition*. Institute for Computer Graphics and Vision, Graz University of Technology, Austria, Technical Report.
9. Buker, U., and G. Hartmann. 2000. Object Representation: On Combining Viewer-centered and Object-Centered Elements. In *International Conference on Pattern Recognition*, vol. 1, 956–959.
10. Ponce, J., T. Berg, M. Everingham, D. Forsyth, M. Hebert, S. Lazebnik, M. Marszałek, C. Schmid, B. Russell, A. Torralba, C. Williams, J. Zhang, and A. Zisserman. 2006. Dataset Issues in Object Recognition. In *Toward Category-Level Object Recognition*, ed. J. Ponce, M. Hebert, C. Schmid, and A. Zisserman. vol. 4170, 29–48. LNCS. Berlin: Springer.
11. Underwood, S., and C. Coates. 1975. Visual Learning from Multiple Views. *IEEE Transactions on Computers* 24 (6): 651–661.
12. Petitjean, S., and B. Loria. 1994. The Complexity and Enumerative Geometry of Aspect Graphs of Smooth Surfaces. In *International Symposium on Effective Methods in Algebraic Geometry*, 143–317.
13. Li, Y., and M.C. Frank. 2007. Computing Non-visibility of Convex Polygonal Facets on the Surface of a Polyhedral CAD Model. *Computer-Aided Design* 39 (9): 732–744.
14. Arif, O., and P. Antonio Vela. 2009. Non-rigid Object Localization and Segmentation using Eigenspace Representation. In *IEEE 12th International Conference on Computer Vision*, 803–808.
15. Wang, Y., and G. De Souza. 2009. A New 3D Representation and Compression Algorithm for Non-rigid Moving Objects Using Affine-Octree. In *IEEE International Conference on Systems, Man and Cybernetics*, 3848–3853.
16. Coelho, E., S. Julier, and B. Maclntyre. 2004. OSGAR: A Scene Graph with Uncertain Transformations. In *IEEE and ACM International Symposium on Mixed and Augmented Reality*, 6–15.

17. Islam, S., D. Silver, and M. Chen. 2007. Volume Splitting and Its Applications. *IEEE Transactions on Visualization and Computer Graphics* 13 (2): 193–203.
18. Bierbaum, A., T. Asfour, and R. Dillmann. 2008. IPSA - Inventor Physics Modeling API for Dynamics Simulation in Manipulation. In *IEEE-RSJ International Conference on Intelligent Robots and Systems, Workshop on Robot Simulators*.
19. Miller, A., and P. Allen. 2004. Graspit! A Versatile Simulator for Robotic Grasping. *IEEE Robotics Automation Magazine* 11 (4): 110–122.
20. Diankov, R. 2010. Automated Construction of Robotic Manipulation Programs. Ph.D. dissertation, Carnegie Mellon University, Robotics Institute.
21. IEEE. 2008. IEEE Standard for Floating-Point Arithmetic - Redline. In *IEEE Std 754-2008 (Revision of IEEE Std 754-1985) - Redline*, 1–82.
22. Cadoli, M., and M. Lenzerini. 1994. The Complexity of Propositional Closed World Reasoning and Circumscription. *Theoretical Computer Science* 48 (2): 255–310.
23. Grimson, W. 1990. *Object Recognition by Computer: The Role of Geometric Constraints*. Cambridge: MIT Press.
24. Fisher, R. 1992. Non-Wildcard Matching Beats The Interpretation Tree. In *British Machine Vision Conference*, 560–569.
25. Shriver, B.D. 1987. Generalizing the Hough Transform to Detect Arbitrary Shapes. In *Readings in Computer Vision: Issues, Problems, Principles, and Paradigms*, ed. M.A. Fischler and O. Firschein, 714–725. USA: Morgan Kaufmann Publishers Inc..
26. Wolfson, H., and I. Rigoutsos. 1997. Geometric Hashing: An Overview. *IEEE Computational Science Engineering* 4 (4): 10–21.
27. Moreno, R., M. Garcia, D. Puig, L. Pizarro, B. Burgeth, and J. Weickert. 2011. On Improving the Efficiency of Tensor Voting. *IEEE Transactions on Pattern Analysis and Machine Intelligence* 33 (11): 2215–2228.
28. Azad, P., T. Asfour, and R. Dillmann. 2009. Accurate Shape-based 6-DoF Pose Estimation of Single-colored Objects. In *IEEE-RSJ International Conference on Intelligent Robots and Systems*, 2690–2695.
29. Murase, H., and S. Nayar. 1993. Learning and Recognition of 3D Objects from Appearance. In *IEEE Workshop on Qualitative Vision*, 39–50.
30. Zhu, K., Y. Wong, W. Lu, and H. Loh. 2010. 3D CAD Model Matching from 2D Local Invariant Fatures. *Computers in Industry* 61 (5): 432–439.
31. Azad, P., D. Muench, T. Asfour, and R. Dillmann. 2011. 6-DoF Model-based Tracking of Arbitrarily Shaped 3D Objects. In *IEEE International Conference on Robotics and Automation*.
32. Roberts, G.. 1963. *Machine Perception of Three-Dimensional Solids*. Outstanding Dissertations in the Computer Sciences. New York: Garland Publishing.
33. Winston, P. 1972. *The MIT Robot*, ed. B. Meltzer. Annual Machine Intelligence Workshop.
34. Guzman, A. 1971. Analysis of Curved Line Drawings using Context and Global Information. In *Machine Intelligence*, ed. B. Meltzer, and D. Mitchie, 325–376. Edinburgh University Press.
35. Agin, G., and T. Binford. 1976. Computer Description of Curved Objects. *IEEE Transactions on Computers* 25 (4): 439–449.
36. Zerroug, M., and R. Nevatia. 1994 . From An Intensity Image to 3-D Segmented Descriptions. In *IAPR-International Conference on Pattern Recognition, Computer Vision Image Processing*, vol. 1, 108–113.
37. Perkins, W. 1978. A Model-Based Vision System for Industrial Parts. *IEEE Transactions on Computers* 27 (2): 126–143.
38. Lowe, D. 1987. Three-Dimensional Object Recognition from Single Two-Dimensional Images. *Artificial Intelligence* 31: 355–395.
39. Song, Y.-Z., B. Xiao, P. Hall, and L. Wang. 2011. In Search of Perceptually Salient Groupings. *IEEE Transactions on Image Processing* 20 (4): 935–947.
40. Sarkar, S. 2003. An Introduction to Perceptual Organization. In *International Conference on Integration of Knowledge Intensive Multi-Agent Systems*, 330–335.
41. Huttenlocher, D.P., and S. Ullman. 1990. Recognizing Solid Objects by Alignment with an Image. *International Journal of Computer Vision* 5: 195–212.

42. Canny, J. 1986. A Computational Approach to Edge Detection. *IEEE Transactions on Pattern Analysis and Machine Intelligence* 8 (6): 679–698.
43. Mundy, J., and A. Heller. 1990. The Evolution and Testing of a Model-based Object Recognition System. In *Third International Conference on Computer Vision*, 268–282.
44. Burns, J.B., R.S. Weiss, and E.M. Riseman. 1992. The Non-existence of General-case View-invariants. In *Geometric Invariance in Computer Vision*, 120–131. Cambridge: MIT Press.
45. Moses, Y., and S. Ullman. 1992. Limitations of Non Model-Based Recognition Schemes. In *European Conference on Computer Vision*, 820–828.
46. Rothwell, C.A. 1995. *Object Recognition Through Invariant Indexing*. Oxford University Science Publications. Oxford: Oxford University Press.
47. Requicha, A.G. 1980. Representations for Rigid Solids: Theory, Methods, and Systems. *ACM Computing Surveys* 12 (4): 437–464.
48. Joshi, S., and T. Chang. 1988. Graph-based Heuristics for Recognition of Machined Features from a 3D Solid Model. *Computer-Aided Design* 20 (2): 58–66.
49. Wu, M., and C. Lit. 1996. Analysis on Machined Feature Recognition Techniques based on B-Representation. *Computer-Aided Design* 28 (8): 603–616.
50. Lowe, D. 2004. Distinctive Image Features from Scale-Invariant Keypoints. *International Journal of Computer Vision* 60 (2): 91–110.
51. Bay, H., T. Tuytelaars, and L.V. Gool. 2006. Surf: Speeded up Robust Features. In *European Conference on Computer Vision*, 404–417.
52. Rothganger, F., S. Lazebnik, C. Schmid, and J. Ponce. 2003. 3D Object Modeling and Recognition using Local Affine-invariant Image Descriptors and Multi-view Spatial Constraints. *IEEE Computer Society Conference on Computer Vision and Pattern Recognition* 66: 272–277.
53. Lepetit, V., L. Vacchetti, D. Thalmann, and P. Fua. 2003. Fully Automated and Stable Registration for Augmented Reality Applications. In *IEEE and ACM International Symposium on Mixed and Augmented Reality*, 93–102.
54. Dementhon, D.F., and L.S. Davis. 1995. Model-based Object Pose in 25 Lines of Code. *International Journal of Computer Vision* 15: 123–141.
55. Hartley, R., and A. Zisserman. 2004. *Multiple View Geometry in Computer Vision*, 2nd edn. Cambridge: Cambridge University Press. ISBN: 0521540518.
56. Fischler, M.A., and R.C. Bolles. 1981. Random Sample Consensus: A Paradigm for Model Fitting with Applications to Image Analysis and Automated Cartography. *Communications ACM* 24 (6): 381–395.
57. Vacchetti, L., V. Lepetit, and P. Fua. 2003. Fusing Online and Offline Information for Stable 3D Tracking in Real-Time. In *IEEE Computer Society Conference on Computer Vision and Pattern Recognition*, vol. 2, 241–248.
58. Kragic, D., A. Miller, and P. Allen. 2001. Real-time Tracking Meets Online Grasp Planning. In *IEEE International Conference on Robotics and Automation*, vol. 3, 2460–2465.
59. Azad, P., T. Asfour, and R. Dillmann. 2009. Combining Harris Interest Points and the SIFT Descriptor for Fast Scale-Invariant Object Recognition. In *IEEE-RSJ International Conference on Intelligent Robots and Systems*, 4275–4280.
60. Lee, S., E. Kim, and Y. Park. 2006. 3D Object Recognition using Multiple Features for Robotic Manipulation. In *IEEE International Conference on Robotics and Automation*, 3768–3774.
61. Piegl, L., and W. Tiller. 1997. *The NURBS Book*. New York: Springer.
62. Lepetit, V., and P. Fua. 2005. Monocular Model-Based 3D Tracking of Rigid Objects: A Survey. *Foundations and Trends in Computer Graphics and Vision* 1 (1): 1–89.
63. Rusu, R., Z. Marton, N. Blodow, M. Dolha, and M. Beetz. 2008. Functional Object Mapping of Kitchen Environments. In *IEEE-RSJ International Conference on Intelligent Robots and Systems*, 3525–3532.
64. Rusu, R., Z. Marton, N. Blodow, A. Holzbach, and M. Beetz. 2009. Model-Based and Learned Semantic Object Labeling in 3D Point Cloud Maps of Kitchen Environments. In *IEEE-RSJ International Conference on Intelligent Robots and Systems*, 3601–3608.

65. Sturm, J., K. Konolige, C. Stachniss, and W. Burgard. 2010. 3D Pose Estimation, Tracking and Model Learning of Articulated Objects from Dense Depth Video using Projected Texture Stereo. In *Workshop RGB-D: Advanced Reasoning with Depth Cameras at Robotics: Science and Systems*.
66. Okada, K., M. Inaba, and H. Inoue. 2003. Walking Navigation System of Humanoid Robot using Stereo Vision based Floor Recognition and Path Planning with Multi-layered Body Image. In *IEEE-RSJ International Conference on Intelligent Robots and Systems*, vol. 3, 2155–2160.
67. Yamaguchi, K., T. Kunii, K. Fujimura, and H. Toriya. 1984. Octree-Related Data Structures and Algorithms. *IEEE Computer Graphics and Applications* 4 (1): 53–59.
68. Balsys, R., D. Harbinson, and K. Suffern. 2012. Visualizing Nonmanifold and Singular Implicit Surfaces with Point Clouds. *IEEE Transactions on Visualization and Computer Graphics* 18 (2): 188–201.
69. Catmull, E., and J. Clark. 1978. Recursively Generated B-spline Surfaces on Arbitrary Topological Meshes. *Computer-Aided Design* 10 (6): 350–355.
70. Marinov, M., and L. Kobbelt. 2005. Optimization Methods for Scattered Data Approximation with Subdivision Surfaces. *Graphical Models* 67 (5): 452–473.
71. Bronshtein, I., K. Semendyayev, G. Musiol, and H. Mühlig. 2003. *Handbook of Mathematics*. Berlin: Springer. ISBN 978-3540721215 .
72. Ben-Arie, J. 1990. The Probabilistic Peaking Effect of Viewed Angles and Distances with Application to 3-D Object Recognition. *IEEE Transactions on Pattern Analysis and Machine Intelligence* 12 (8): 760–774.
73. Olson, C. 1994. Probabilistic Indexing: A New Method of Indexing 3D Model Data from 2D Image Data. In *CAD-Based Vision Workshop*, 2–8.
74. Li, W., G. Bebis, and N. Bourbakis. 2008. 3-D Object Recognition using 2-D Views. *IEEE Transactions on Image Processing* 17 (11): 2236–2255.
75. Bohannon, P., P. Mcllroy, and R. Rastogi. 2001. Main-memory Index Structures with Fixed-size Partial Keys. *SIGMOD International Conference on Management of data* 30 (2): 163–174.
76. Duda, R., P. Hart, and D. Stork. 2001. *Pattern Classification*, 2nd edn. New York: Wiley. ISBN 978-0471056690.
77. Härdle, W., M. Müller, S. Sperlich, and A. Werwatz. 2004. *Nonparametric and Semiparametric Models*. New York: Springer.
78. Khronos Group. 2012. COLLADA - Digital Asset and FX Exchange Schema. https://collada.org/.
79. Requicha, A. 1977. Mathematical Models of Rigid Solid Objects. Production Automation Project, University of Rochester, Technical Report 28.
80. Przybylski, M., T. Asfour, and R. Dillmann. 2011. Planning Grasps for Robotic Hands using a Novel Object Representation based on the Medial Axis Transform. In *IEEE-RSJ International Conference on Intelligent Robots and Systems*, 1781–1788.
81. Blodow, N., C. Goron, Z. Marton, D. Pangercic, T. Rühr, M. Tenorth, and M. Beetz. 2011. Autonomous Semantic Mapping for Robots Performing Everyday Manipulation Tasks in Kitchen Environments. In *IEEE-RSJ International Conference on Intelligent Robots and Systems*, 4263–4270.
82. Baresi, L., E.D. Nitto, and C. Ghezzi. 2006. Towards Open-World Software: Issue and Challenges. In *Annual IEEE/NASA Software Engineering Workshop*, 249–252.
83. Yamauchi, B., and R. Beer. 1996. Spatial Learning for Navigation in Dynamic Environments. *IEEE Transactions on Systems, Man, and Cybernetics, Part B: Cybernetics* 26 (3): 496–505.
84. Brooks, R. 1981. Symbolic Reasoning Among 3-D Models and 2-D Images. *Artificial Intelligence* 17 (3): 285–348.
85. Burgess, N. 2006. Spatial Memory: How Egocentric and Allocentric Combine. *Trends in Cognitive Sciences* 10 (12): 551–557.

Chapter 4
Methods for Robust and Accurate Image Acquisition

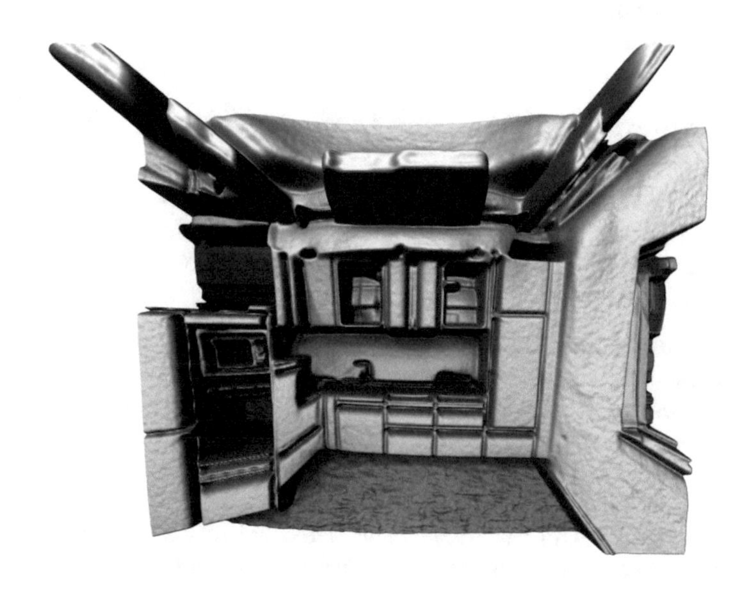

4.1 Visual Manifold

The visual perception system of a humanoid robot should be able to attain and manage the vision model coupling. This essential link between the physical reality and the world model representation has to be supported by diverse visual skills. The efficiency, robustness and precision of these skills directly depend on the extracted visual-features. The amount, representativeness and repeatability of the visual-features rely upon the stability, accuracy and dynamic range of the sensed visual manifold.

Therefore, in order to properly capture the visual manifold for vision model coupling, two complementary methods are presented:

D. I. González Aguirre, *Visual Perception for Humanoid Robots*,
Cognitive Systems Monographs 38, https://doi.org/10.1007/978-3-319-97841-3_4

- **Optimal irradiance signals**: A novel method is introduced for consistent and stable image acquisition based on multiple image fusion. The method reliably captures the visual manifold by optimally estimating the sensor irradiance signals.
- **Expanding dynamic range**: Humanoid robots are confronted with high-contrast scenes, dynamically varying illuminations and complex surface materials. The wide intra-scene radiance is usually beyond the standard dynamic range of the camera sensors. A novel method for synthesizing accurate high dynamic range (HDR) images for complex heterogeneously lighted scenes is presented to enable humanoid robots to cope with real environmental conditions

Since both methods are based on multiple image fusion, their synergistic combination acquires the visual manifold with a high signal-to-noise ratio and high dynamic range. The experimental evaluation (in Sects. 4.6 and 4.8) corroborates the quality and suitability of the methods.

4.2 Hardware Limitations and Environmental Conditions

The particular morphology of a humanoid robot allows various advantages for their application (see details in Chap. 1). However, this fact also imposes restrictions on their effectors and sensors. Specifically, the visual perception through stereoscopic vision is restricted by:

- **Physical constraints**: These conditions restrain the stereo base-line length, size and weight of the cameras and lenses (see Fig. 4.1).
- **Complex perturbations**: Inside a humanoid robot several devices coincide. The simultaneous operation of sensors and actuators produces electric, magnetic, mechanical and thermal perturbations which deteriorate the quality of the sensor signals (see Fig. 4.2).
- **Circumscribed resources**: Autonomous humanoid robots have to dependably execute complex tasks supported by limited memory and computational power available on board. Early approaches using remote computation (outside the humanoid) have been investigated, for instance in [1]. This paradigm limited the autonomy and application potential of humanoid robots.
- **Extensive requirements**: The resources for visual sensing skills change dynamically, for example, the resolution, frame rate and geometric configuration of the cameras for vergence, smooth-pursuit, tracking [2] and saccades [3].

Even under these conditions, the visual perception system of a humanoid robot requires consistent and stable sensor observations in order to attain the vision model coupling. A multiple image fusion method based on density estimation is introduced in Sect. 4.5 for overcoming these limitations. The stability and robustness of this method cope with arbitrary multimodal distributions of the irradiance signals.

The proposed fusion method is suitable for both gray scale and color images using color filter arrays such as the wide spread Bayer pattern [6]. The resulting

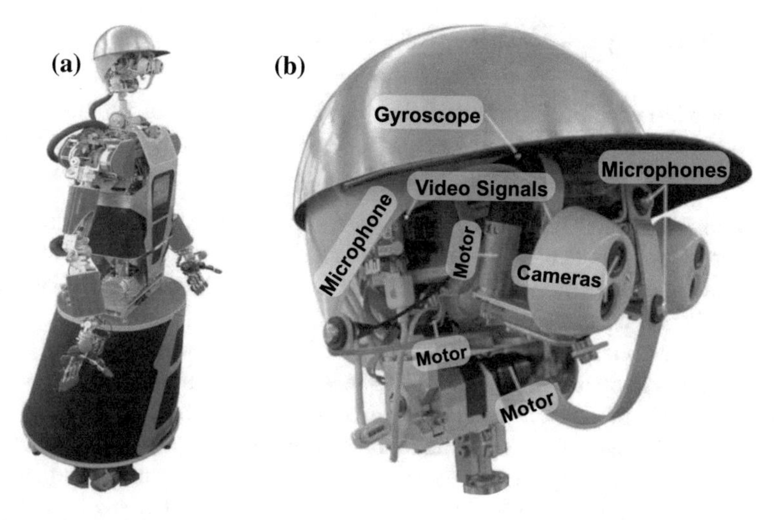

Fig. 4.1 Implementation constraints to be considered for integrated image acquisition processes of a humanoid robot. **a** The humanoid robot ARMAR-IIIa [4]. **b** This humanoid robot is equipped with four cameras, six motors, a gyroscope and six microphones [5]. During task execution, these devices simultaneously work in a perception-planning-action cycle

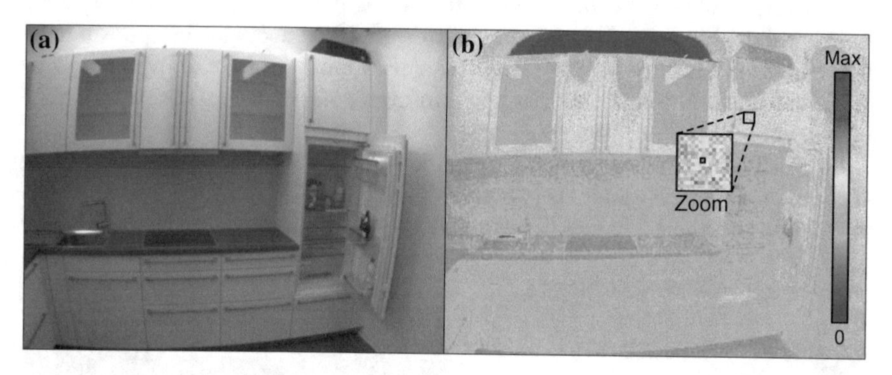

Fig. 4.2 Environmental difficulties to be considered for robust image acquisition for humanoid robots. **a** An everyday scene in a made-for-humans environment where humanoid robots have to recognize objects and estimate their positions and orientations. **b** The pseudo-color deviation image shows the camera instability and detrimental artifacts produced from different noise sources

fused images improve the global stability and precision of the visual perception processes (see experimental evaluation in Sect. 4.6). Furthermore, the performance of the method allows the acquisition of semi-dynamic[1] scenes in real applications.

[1]The content of the scene and the robot remain static during visual sensing.

4.3 Visual Acquisition: Related Work

There is a wide variety of methods for improving the quality of captured images. These methods can be split in two main categories; image rectification through enhancement and image synthesis by fusion.

Image enhancement: The methods in this category deal with the inverse problem of estimating an optimal *"noiseless"* image from a single noisy image. For more than four decades, image denoising and image enhancement algorithms have achieved considerable progress in image restoration.

Most image enhancement methods are based on rank and neighborhood filters (see extensive survey in [7]). For instance, the k-nearest neighbors [8] and non-local means filters [9–11] provide considerable results without blotchy artifacts when using small windows. Their computational performance (up to 500 FPS using a GPU as reported in [12]) decreases quadratically depending on the applied window. Nevertheless, when noise affects a region beyond one or two pixels the resulting images show severe blotchy artifacts. This occurs because the perturbations are arbitrarily distributed with non-causal nature in terms of intensity variations, in particular Fig. 4.10). Thus, the robustness of these methods is not adequate for everyday humanoid robot applications. Recently, the method in [13] showed outstanding results with highly contaminated images. However, that approach has proven to be computationally expensive. The method requires more than 3 min even with low resolution (256x256 pixels) images. Its performance makes it prohibitive for online applications.

In general, all image enhancement methods can only improve the image up to so-called *"one-image"* limit. This occurs due to diminishing factors which cannot be filtered from a single image. For example, the flickering produced by the artificial

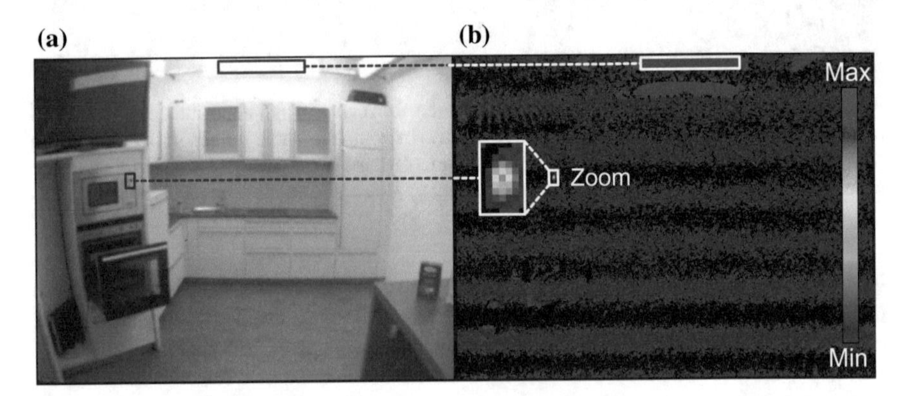

Fig. 4.3 Combined environmental and embodiment sensing perturbations. **a** Everyday scene in a household environment. **b** The deviation of a single image shows important aspects: (i) Noise waves disturb (partially vanish or even completely occlude) the underlying saliences such as edges and corners. (ii) The inadequately (under or over) exposed regions expose no deviation

lighting in an indoor environment. Even more important, within humanoid robots with a high number of active joints, the electromagnetic perturbations produced by the motors in the head (yaw, pitch, roll) generate noise waves on the image (see Fig. 4.3). Due to the variable payload of the head motors, this non-parametric distributed image noise depends on the kinematic configuration of the humanoid robot head. In such situations, a single image locally loses information in the affected region and the noiseless-inverse extraction cannot be soundly and efficiently solved with existing methods.

Image fusion: These methods exploit the available information in two or more images in order to synthesize a new improved image. Image fusion has been widely investigated and successfully applied in various domains (see survey [14]). Depending on the nature of the data and the intended results, these methods are categorically divided into:

- **Image registration**: Transforms image sets captured from different viewpoints and sensors into a consistent coordinate system [15].
- **Super resolution**: Enhances the image resolution by means of structural similarity [16] or statistical regularity [17].
- **High dynamic range**:Increases the irradiance range beyond the sensor capabilities by fusing differently exposed images (see Sect. 4.7).
- **Multiple focus imaging**:Expands the depth of field by fusing images with various focal configurations [18].
- **Poisson blending**:Composes gradient domain images for the seamless image content insertion [19].
- **Image based rendering**:Uses multiple images to generate novel scene viewpoints [20].
- **Image Stitching**:Combines images with overlapping field of view in order to produce panoramas [21].

In contrast to these methods, the aim of the first method proposed in this chapter is to improve the quality and stability of the images acquired in each viewpoint using the available sensor resolution. The key idea of the proposed method is to observe the scene by capturing and fusing various images instead of restoring a single contaminated image.

In terms of category, the proposed method is related to image registration and image-based rendering. However, the aim of the method is not to generate images from different viewpoints but to extract the optimal image from a set of noisy images.

In order to support humanoid robots in everyday applications, the method considers the following aspects:

- **Applicability**: No assumptions regarding image content is done. This enables wide range of applicability for humanoid robots.
- **Quasi-dynamic**: The robot and the scene are static while the observations place. For applicability purposes, this should take less than one second (see Sect. 4.6.5).
- **Fixed calibration**: Geometric consistency is well-kept for extraction of Euclidean metric from stereo images. The fixed focal length is also preserved according to the intrinsic camera calibrations (see stereo focusing in [22]).
- **Visual-feature metrics**: The stability and quality of the acquired images are improved and measured in terms of image processing results, not in human perceptual metrics.

4.4 Visual Manifold Acquisition Pipeline

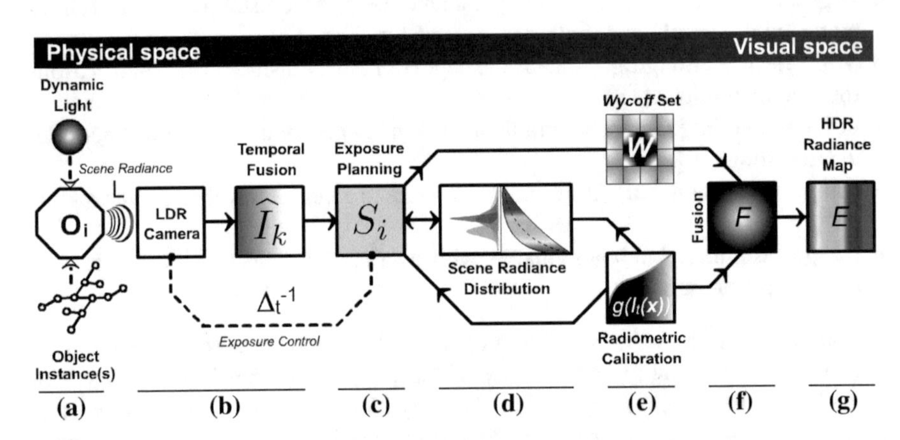

Fig. 4.4 The visual manifold acquisition pipeline. **a** Complex high-contrast heterogeneously lighted scene (for instance the scene in Fig. 4.16). **b** The low dynamic range (LDR) cameras capture differently exposed sets of images, called *Wycoffset* in [23]. Those image sets are used for temporal fusion to obtain *"noiseless"*, images $\widehat{I_k^i}$ (see Sect. 4.5). **c** The exposure planning determines the amount and integration time Δt_i of each of the images to optimally sample the scene radiance. **d** The radiance distribution of the scene is captured by the minimal amount of exposures $\mathcal{E}_{\min}$ covering the full reachable radiance by the camera sensor. This ensures the minimal sampling density. **e** The radiometric calibration is the response function of the composed optical system. **f** The Wycoffset set and the radiometric calibration are used to synthesize $[F]$ the HDR image. **g** The resulting HDR image E is the coherent and precise (up-to-scale) visual manifold

4.5 Temporal Image Fusion

The first step for robust image acquisition by multiple image fusion is to analyze the sensor deviation behavior (described in Sect. 4.5.1). Based on this behavior, an efficient and robust fusion strategy is introduced to overcome both the sensor deficiencies and the unsuitable environmental circumstances (in Sect. 4.5.5). Afterwards, the convergence analysis provides a deeper insight into the selection and effects of the amount of images to fuse, namely the sampling horizon (in Sect. 4.5.9). Thereafter (in Sect. 4.6), two fundamental feature extractions are computed in order to evaluate the stability and precision improvements of the method. Subsequently, the relation between exposure (integration time) and convergence of the fusion process is analyzed in order to determine the minimal necessary amount of images to fuse while ensuring the resulting image quality. This relation is essential to keep the amount of images practicable (see Sect. 4.6.4) while expanding the dynamic range (in Sect. 4.7).

4.5.1 Sensing Deviation

Within a semi-dynamic scene, both the robot and the elements in the field of view remain static for a short period of time (at most 1 sec). However, the illumination conditions can change dynamically. For instance, the flickering of the lamps or the effects of natural illumination variations from reflections, refractions and shadows produced by windows or other environmental conditions. Assuming a semi-dynamic scene, the sensing error is proportional to the amount of intensity variation. In order to anticipate possible sampling artifacts[2] and reduce the total sensing time during the capture of multiple images, the maximal frame rate is used while the images are indirectly stored for optimal image estimation (see Sect. 4.5.7).

4.5.2 Irradiance Signals

The location $\mathbf{x} \in \mathbb{N}^2$ within the pixel location set $\Omega \subset \mathbb{N}^2$ is limited by the width $w \in \mathbb{N}$ and height $h \in \mathbb{N}$ of the image. The discrete intensity set $\Theta \in \mathbb{N}$ is the codomain of the time varying image function $I_t : \mathbb{N}^2 \mapsto \mathbb{N}$. The value associated with each pixel location $\mathbf{x}$ is a random variable independent and identically distributed over the intensity set Θ as

$$\mathbf{x} \in \Omega := \{\, x \mid (1, 1) \leq x \leq (w, h) \,\} \subset \mathbb{N}^2, \tag{4.1}$$

$$I_t(\mathbf{x}) \in \Theta := \left\{\, i \mid 0 \leq i \leq (2^m - 1) \,\right\} \subset \mathbb{N}, \tag{4.2}$$

where t is the time stamp and m denotes the (usually 8) bits per pixel.

---

[2]According to the *Nyquist-Shannon* sampling theorem [24].

4.5.3 Temporal Scope

Because the sensing process is realized by fusing multiple images of the scene, the concept of temporal scope is a general way to formulate the following method without specific camera considerations. The observation time scope includes $k > 1$ images. There are various descriptive statistics which provide the initial means for analyzing the pixel irradiance signal $I_t(\mathbf{x})$. The maximum value within the k temporal scope is denoted as

$$\mathbf{U}_k(\mathbf{x}) := \mathbf{max}\Big[I_t(\mathbf{x})\Big]_{t=1}^{k}. \tag{4.3}$$

In the same manner the minimal value is expressed as

$$\mathbf{L}_k(\mathbf{x}) := \mathbf{min}\Big[I_t(\mathbf{x})\Big]_{t=1}^{k}. \tag{4.4}$$

These upper and lower intensity boundaries describe the variation range within this temporal scope as

$$\mathbf{R}_k(\mathbf{x}) := \mathbf{U}_k(\mathbf{x}) - \mathbf{L}_k(\mathbf{x}). \tag{4.5}$$

Finally, the mean intensity of the scope is denoted by

$$\mathbf{M}_k(\mathbf{x}) := \frac{1}{k}\sum_{t=1}^{k} I_t(\mathbf{x}). \tag{4.6}$$

These basic descriptive statistics are helpful to identify distinct essential phenomena (see Fig. 4.5). First, the intensity signal significantly and arbitrarily varies from one frame to another without deterministic pattern. Second, the intensity range (Eq. 4.5) increases rapidly (in the first 10 frames) followed by a much slower increasing rate which remains unchanged after 100 to 120 frames. Third, the mean value (Eq. 4.6) converges fast at the beginning of the sampling (approximately after 30 frames). Finally, the mean value keeps converging in a very slow rate.

4.5.4 Central Limit

These observations could lead to a straightforward fusion strategy using the mean value $\mathbf{M}_k$ within a time scope of approximately one second. This will also follow the intuition behind the *central limit* theorem. This theorem states: *The mean of a sufficiently large number of independent random variables with finite mean and variance* (which is the case of the signal I_t) *will be approximately normally distributed* [25].

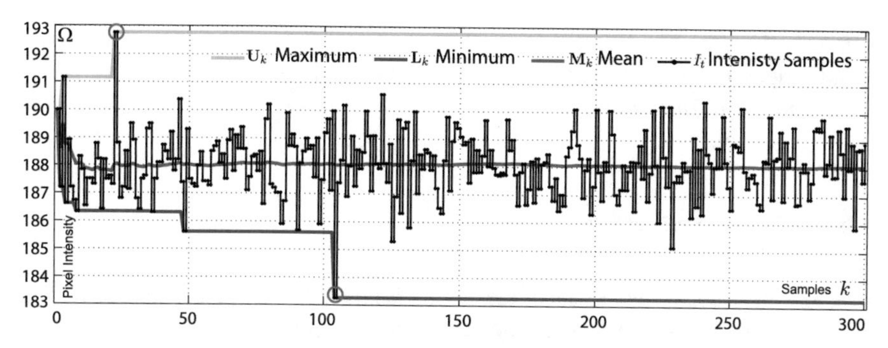

Fig. 4.5 The intensity signal of a pixel $I_t(\mathbf{x})$ along the k temporal scope. The sequential intensity values are used to show the descriptive statistics (from Eqs. 4.3 to 4.6). The magenta circles denote the highest and lowest intensity outliers. The pixel source location is marked in Fig. 4.2

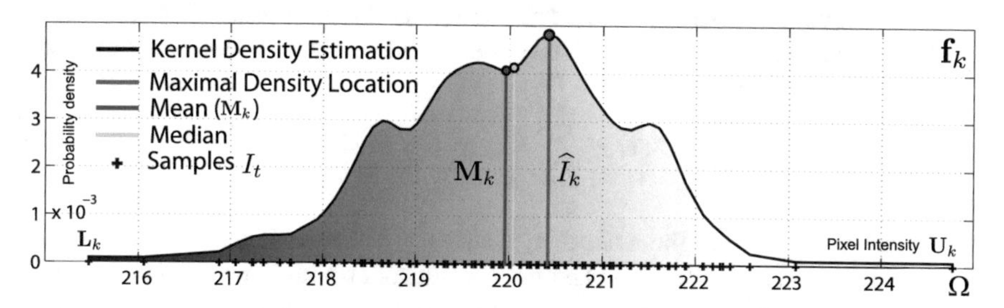

Fig. 4.6 The probabilistic distribution function (Eq. 4.7) clearly illustrates that the straightforward fusion through mean or median can produce suboptimal fusion density. This occurs when the intensity samples are not symmetrically spread, slightly skewed or multimodal distributed

However, this theorem cannot be soundly applied to the limited amount of samples (captured in one second) in order to model them with a normal distribution. This occurs because two criteria are not *"sufficiently"* fulfilled, namely:

- **Approximately normal**: For each pixel, a single independent random variable is modeled. Thus, the distribution can exhibit rather large skewness and is (usually) neither symmetric nor unimodal. A discussion whether the nature of each pixel signal arises from a single or a complex process (with a large number of small effects) can still support the use of generalizations of the central limit theorem. However, the quality of the approximation using a reduced amount of samples would not be appropriate (see Fig. 4.6).
- **Asymptotic convergence**: The convergence of the central limit theorem can only be reached with a rather large amount of samples. Actually, it is only a tendency to be reached at infinity. The speed of convergence is determined (under all previous considerations) by the *Berry-Esseen* theorem [25] as $C = k^{-\frac{1}{2}}$, where $C \in \mathbb{R}$ is a scalar denoting the rate of convergence and $k \in \mathbb{N}^+$ is the number of images in the

temporal scope. The Berry-Esseen theorem provides a useful convergence speed but it does not reveal the convergence level which is actually the critical quality indicator. This theorem uprises experimentally (in Sect. 4.5.9).

4.5.5 Fusion Strategy

Based on previous observations of the central limit theorem, in order to fuse various images the probability density function (PDF) of each pixel can be used to determine the representativeness of the captured samples. In this manner, no assumptions about the distribution are held. This makes the intensity estimation process general and robust. The continuous PDF can be attained by KDE (Kernel Density Estimation [26]). This approach provides advantages compared to the straightforward mean or median fusion (see Fig. 4.6). The KDE of the irradiance signal (Eq. 4.2) is expressed as

$$\mathbf{f}_k(\mathbf{x}, i) = \sum_{t=1}^{k} K_{\lambda_k(\mathbf{x})}\Big(i - I_t(\mathbf{x})\Big), \tag{4.7}$$

where the $K_{\lambda_k(\mathbf{x})}$ denotes the smoothing Gaussian kernel with pixel-wise adaptive bandwidth $\lambda_k(\mathbf{x})$. Notice that the bandwidth assessment is an issue itself [27]. However, in the one dimensional case the Silverman rule [28] is used to efficiently estimate the pixel-wise bandwidth.

In contrast to ϵ-truncation or kernel dependency [27], the PDF is efficiently approximated by this adaptive method without accuracy drawbacks. Moreover, in compliance with the central limit theorem, the most likely fusion value $\widehat{I_k} \in \mathbb{R}$ is the intensity location with the highest probability density within the temporal scope (the top red location in Fig. 4.6) expressed as

$$\widehat{I_k}(\mathbf{x}) := \mathbf{argmax}_{i \in \mathbb{R}}\, \mathbf{f}_k(\mathbf{x}, i). \tag{4.8}$$

Substituting Eq. 4.7 in Eq. 4.8 and considering the upper intensity (Eq. 4.3) and lower intensity (Eq. 4.4) boundaries of the k sampling scope, the fusion objective function can be written as

$$\widehat{I_k}(\mathbf{x}) := \mathbf{argmax}_{i \in \mathbb{R}} \left[\sum_{t=1}^{k} K_{\lambda_k(\mathbf{x})}\Big(i - I_t(\mathbf{x})\Big)\right]_{i=\mathbf{L}_k(\mathbf{x})}^{\mathbf{U}_k(\mathbf{x})}. \tag{4.9}$$

Equation 4.9 is the bounded description of the optimization problem for each pixel. This formulation does not only attain the most likely intensity value (intensity location with highest probabilistic density) but it also improves the resulting nature of the fusion image, from a noisy discrete collection of image samples $I_t : \mathbb{N}^2 \mapsto \mathbb{N}$ to a single *"noiseless"* and semicontinuous range function $\widehat{I_k} : \mathbb{N}^2 \mapsto \mathbb{R}$. This optimization plays an important role for both the increasing of image dynamic range (Sect. 4.7) and improving visual-feature extraction (Sect. 5.6). The straightforward optimization (Eq. 4.9) with high precision requires minutes to be computed using the onboard CPUs of the humanoid robot ARMAR-IIIa. In order to efficiently solve the pixel-wise optimization problem in less than a second, the following data structure and two-phase execution method are proposed.

4.5.6 Fusion Data Structure

The fusion structure $Z(\mathbf{x}) \in \mathbb{N}^{[2^m+5]}$ (recall m stands for bit depth in Eq. 4.2) consists of a six-tuple per pixel

$$Z(\mathbf{x}) := \left\langle H_{\mathbf{x}},\ \mathbf{U}_k,\ \mathbf{L}_k,\ \mathbf{Acc}_k,\ \mathbf{Asc}_k,\ \mathbf{f}_{k_{\text{Max}}} \right\rangle, \tag{4.10}$$

were the pixel histograms $H_{\mathbf{x}} \in \mathbb{N}^{2^m}$, maximal intensity $\mathbf{U}_k(\mathbf{x})$ (Eq. 4.3), minimal intensity $\mathbf{L}_k(\mathbf{x})$ (Eq. 4.4), intensity accumulator

$$\mathbf{Acc}_k(\mathbf{x}) := \sum_{t=1}^{k} I_t(\mathbf{x}), \tag{4.11}$$

square intensity accumulator

$$\mathbf{Asc}_k(\mathbf{x}) := \sum_{t=1}^{k} I_t(\mathbf{x})^2 \tag{4.12}$$

and the maximal bin frequency $\mathbf{f}_{k_{\text{Max}}} \in \mathbb{N}$ are stored per pixel. Based on the dimensions of the image (defined in Eq. 4.1), the amount of memory required to store six-tuples $Z(\mathbf{x})$ for all pixels is $(2^{[m+1]} + 16)wh$. This six-tuples occupy a rather large memory space because there is a histogram per pixel allocating two bytes per bin in order to enable the fusion of large sets of images. This feature is needed during the convergence trade-off analysis (in Sect. 4.5.10). In the experimental evaluation, the space required to store this structure for an image of 640×480 pixels and 8 bits of depth is 528 bytes per pixel and almost 155 MB for the whole image.

4.5.7 Sensing Integration

After capturing one of the $t \leq k$ images I_t, the sensed intensity values are incorporated into the fusion data structure (see Algorithm 1). Since these operations are performed using the aligned[3] data structure (Eq. 4.10), the sensing integration requires less time than the image transmission from the camera to the CPU even at 60 FPS. Hence, the sensing integration stage produces no delay in the sensing process. The advantage of the inter-frame processing is the use of sensor integration and transmission time to load the information from the previous frame. The six-tuple also has algorithmic advantages resulting from inter-frame processing: (i) The mean intensity and standard deviation are computed without additional iterations. (ii) The intensity range (Eq. 4.5) and maximal histogram frequency $\mathbf{f}_{k\text{Max}}$ of all pixels are available directly after the sensing. (iii) The histogram approximates the irradiance PDF for the continuous KDE estimation to be efficiently computed (see Algorithm 3).

Algorithm 1 Sensing-Integration, $SI(k, I_t, Z, \Delta t_j)$

Require: $k > 1$
1: {Reset: Clean the data structure Z}
2: **for** $\mathbf{x}$ in Ω **do**
3: $\quad H_\mathbf{x} \leftarrow [0]_{2^m}$, $\mathbf{U}_k(\mathbf{x}) \leftarrow -\infty$, $\mathbf{L}_k(\mathbf{x}) \leftarrow \infty$
4: $\quad \mathbf{Acc}_k(\mathbf{x}) \leftarrow 0$, $\mathbf{Asc}_k(\mathbf{x}) \leftarrow 0$, $\mathbf{f}_{k\text{Max}} \leftarrow 0$
5: **end for**
6: {Sensing Integration: Capture and incorporate the irradiance signals}
7: $I_1 \leftarrow$ *ImageCapture*(Δt_j);
8: **for** $t = 1$ **to** k **do**
9: $\quad$ {Iso-exposed Δt_j capturing: See exposure in Sect. 4.7.6}
10: $\quad$ *BeginImageCapture*(Δt_j);
11: $\quad$ **for** $\mathbf{x}$ in Ω **do**
12: $\qquad H_\mathbf{x}[I_t(\mathbf{x})] \leftarrow H_\mathbf{x}[I_t(\mathbf{x})] + 1$, $\mathbf{f}_{k\text{Max}} \leftarrow \mathbf{max}(H_\mathbf{x}[I_t(\mathbf{x})], \mathbf{f}_{k\text{Max}})$
13: $\qquad \mathbf{U}_k(\mathbf{x}) \leftarrow \mathbf{max}(I_t(\mathbf{x}), \mathbf{U}_k(\mathbf{x}))$, $\mathbf{L}_k(\mathbf{x}) \leftarrow \mathbf{min}(I_t(\mathbf{x}), \mathbf{L}_k(\mathbf{x}))$
14: $\qquad \mathbf{Acc}_k(\mathbf{x}) \leftarrow \mathbf{Acc}_k(\mathbf{x}) + I_t(\mathbf{x})$, $\mathbf{Asc}_k(\mathbf{x}) \leftarrow \mathbf{Asc}_k(\mathbf{x}) + I_t(\mathbf{x})^2$
15: $\quad$ **end for**
16: $\quad I_{t+1} \leftarrow$ *EndImageCapture*();
17: **end for**

4.5.8 *Iso-Exposed Fusion: Optimal Intensity Estimation*

Once the sensing integration is completed for all k identically exposed images, the approximation PDF of each pixel $\mathbf{x}$ is available in the corresponding histogram $H_\mathbf{x}$. The continuous estimation of the PDF (Eq. 4.7) can profit from this discrete PDF in order to determine the regions of the intensity domain (Eq. 4.9) where the maximal density is located. This prevents unprofitable KDE computations in low likelihood regions. The high density regions are extracted in the pixel-wise interval set $\mathbf{W}_\mathbf{x}$

$$\mathbf{W}_\mathbf{x} := \left\{ \Big[\mathbf{max}(j - \lambda_k(\mathbf{x}), \mathbf{L}_k), \mathbf{min}(j + \lambda_k(\mathbf{x}), \mathbf{U}_k) \Big] \mid \mathbf{L}_k \leq j \leq \mathbf{U}_k \right\}, \quad (4.13)$$

[3]Algorithmically, this is a common trade-off between space and time complexity. Technically, this continuous large memory block reduces the processing time.

where the adaptive bandwidth λ_k is estimated by Silverman rule [28]

$$\lambda_k(\mathbf{x}) = \left(\frac{4\sigma_k(\mathbf{x})^5}{3k}\right)^{\frac{1}{5}}, \tag{4.14}$$

which depends on the standard deviation (optimally computed) as

$$\sigma_k^2(\mathbf{x}) = \mathbf{Asc}_k(\mathbf{x}) - \left(\frac{1}{k}\mathbf{Acc}_k(\mathbf{x})\right)^2. \tag{4.15}$$

Notice that Eq. 4.13 is subject to $\forall H_{\mathbf{x}}[j] \geq \lfloor \frac{1}{2}\mathbf{f}_{k\text{Max}} \rfloor$ ensuring intervals in the higher quartile and

$$H_{\mathbf{x}}\Big[\mathbf{max}(j-1, \mathbf{L}_k)\Big] \leq H_{\mathbf{x}}[j] \geq H_{\mathbf{x}}\Big[\mathbf{min}(j+1, \mathbf{U}_k)\Big], \tag{4.16}$$

for guaranteeing the optimization search within the local maxima or (at least due to discretization effects) in the high density plateaus of the PDF. The interval set $\mathbf{W}_{\mathbf{x}}$ is efficiently computed (in Algorithm 2) using the descriptive statistics within the fusion structure $Z(\mathbf{x})$ (see Eq. 4.10).

Algorithm 2 Extract-Interval-Set, $EIS(\lambda_k, H_{\mathbf{x}}, \mathbf{L}_k, \mathbf{U}_k, \mathbf{f}_{k\text{Max}})$

Require: $\mathbf{L}_k \neq \mathbf{U}_k$ **and** $\lambda_k \geq 1$
1: $\mathbf{W}_{\mathbf{x}} \leftarrow \emptyset$
2: **for** $j_0 = \mathbf{L}_k$ **to** $\mathbf{U}_k$ **do**
3: **if** $(H_{\mathbf{x}}[j_0] \geq \lfloor \frac{1}{2}\mathbf{f}_{k\text{Max}} \rfloor)$ **then**
4: **if** $(H_{\mathbf{x}}[j_0] \geq H_{\mathbf{x}}[\mathbf{max}(j_0 - 1, \mathbf{L}_k)])$ **then**
5: $j_1 \leftarrow \mathbf{min}(j_0 + 1, \mathbf{U}_k)$
6: **while** $((j_1 \leq \mathbf{U}_k) \wedge (H_{\mathbf{x}}[j_1] \geq H_{\mathbf{x}}[\mathbf{min}(j_1 + 1, \mathbf{U}_k)]))$ **do**
7: $j_1 \leftarrow j_1 + 1$
8: **end while**
9: $\mathbf{W}_{\mathbf{x}} \leftarrow \mathbf{W}_{\mathbf{x}} \cup [\mathbf{max}(j_0 - \lambda_k, \mathbf{L}_k), \mathbf{min}(j_1 + \lambda_k, \mathbf{U}_k)]$
10: **end if**
11: **end if**
12: **end for**
13: **return** $\mathbf{W}_{\mathbf{x}}$

Singularly in the presence of distant outliers, the narrow and reliable search space contained in the interval set $\mathbf{W}_{\mathbf{x}}$ reduces the computations in a notable manner. This occurs because the KDE is only coarsely and evenly computed with increments $0 < \alpha < \frac{1}{5}$ in the intervals $l \in \mathbf{W}_{\mathbf{x}}$ as

$$\widehat{I_k^{\alpha}}(\mathbf{x}) := \mathbf{argmax}_{i \in \mathbb{R},\ l \in \mathbf{W}_{\mathbf{x}}} \left[\sum_{t=1}^{k} K_{\lambda_k(\mathbf{x})}\Big(i - I_t(\mathbf{x})\Big)\right]_{i=\alpha \cdot r + \triangleright(l),\ r \in \mathbb{N}_0}^{\triangleleft(l)}, \tag{4.17}$$

where the operators ($\triangleleft, \triangleright$) denote the left and right boundaries of the interval element l from the set $\mathbf{W_x}$. While computing Eq. 4.17, the complexity is implicitly adjusted according to the observed intensity range and distribution of the local maxima. The pixel histogram $H_\mathbf{x}$ also reduces the KDE computational complexity. When accumulating the weighting contribution of each sample (line 3 in Algorithm 3), the expensive Gaussian kernel is computed only once for all samples within the intensity bin.

Algorithm 3 Histogram-Based-KDE, $HB\text{-}KDE(i,\lambda_k,H_\mathbf{x},\mathbf{L}_k,\mathbf{U}_k)$

Require: $\lambda_k \geq 1$
1: $\delta_i \leftarrow 0,\ j_0 \leftarrow \mathbf{max}(\lfloor i - \lambda_k \rfloor, \mathbf{L}_k),\ j_1 \leftarrow \mathbf{min}(\lceil i + \lambda_k \rceil, \mathbf{U}_k)$
2: **for** $j = j_0$ **to** j_1 **do**
3: $\quad \delta_i \leftarrow \delta_i + \underbrace{(K_{\lambda_k}(i-j) \cdot H_\mathbf{x}[j])}_{\text{adaptive complexity reduction}}$
4: **end for**
5: **return** δ_i

The value precision of $\widehat{I}_k^\alpha(\mathbf{x})$ is controlled by the increment α (see Eq. 4.17). When higher precision ($\beta \ll \alpha$) is intended a rather small β has to be used. This implies high computational workload with a scene and pixel-wise variable complexity of $O_\mathbf{x}(\mathbf{R}_k(\mathbf{x})/\beta)$. In order to prevent this, an approximation can be obtained first by a scattered α-sampling and subsequently followed by a β-refinement $\widehat{I}_k^\beta(\mathbf{x})$ within the convex interval of the global maxima. This auto-adjustment (lines 25–36 in Algorithm 4) reduces the computational complexity of the refinement to $O\,(log_2\,(\alpha/\beta))$. These two phases sagaciously adapt the required workload while obtaining higher accuracy than a dense sampling approach (Eq. 4.9). Finally, the complete optimization is presented (in Algorithm 4) followed by qualitative results (in Fig. 4.7).

4.5.9 Convergence Behavior

Based on the descriptive statistics, it is obvious that the stability of the sensor values is reached (at latest) when the global range increment is neglectable. This behavior is globally and smoothly depicted by the range expansion $\mathbf{E}_k := \frac{1}{wh}\sum_{\mathbf{x}\in\Omega}\mathbf{R}_k(\mathbf{x})$ and its rate $\psi_k := \frac{\delta \mathbf{E}_k}{\delta k}$.

Usually, the range expansion $\mathbf{E}_k$ requires a large temporal scope k in order to converge (see Fig. 4.8). This slow convergence occurs due to the outliers in the noisy lower and upper quantiles (for instance, the samples at frames 20 and 104 in Fig. 4.5). The range expansion rate ψ_k provides a shorter limit to estimate the number of images needed for optimal fusion. However, it is observable that after fusing certain number of images the resulting image does not improve substantially. In order to determine the minimal number of images required for the convergence, the following approach was realized. First, a large image set is captured. Its cardinality is called n-horizon. Using this set, the fusion strategy is performed (through Eq. 4.17). In subsequent

Algorithm 4 Optimal-Intensity-Estimation, $OIE(k, \alpha, \beta, Z, \widehat{I_k})$

Require: $k > 1,\ 0 < \alpha \leq \frac{1}{5}$ **and** $0 < \beta \ll \alpha$

1: **for** $\mathbf{x}$ in Ω **do**
2: **if** $(\mathbf{L}_k(\mathbf{x}) = \mathbf{U}_k(\mathbf{x}))$ **then**
3:   $\widehat{I_k}(\mathbf{x}) \leftarrow \mathbf{U}_k(\mathbf{x})$
4: **else**
5:   $\mathbf{M}_k(\mathbf{x}) \leftarrow \frac{1}{k}\mathbf{Acc}_k(\mathbf{x}),\ \sigma_k(\mathbf{x}) \leftarrow (\mathbf{Asc}_k(\mathbf{x}) - \mathbf{M}_k(\mathbf{x})^2)^{\frac{1}{2}}$.
6:   $\lambda_k(\mathbf{x}) \leftarrow (\frac{4}{3k}\sigma_k(\mathbf{x})^5)^{\frac{1}{5}}$
7:   $\mathbf{W}_\mathbf{x} \leftarrow EIS(\lambda_k(\mathbf{x}), H_\mathbf{x}, \mathbf{L}_k(\mathbf{x}), \mathbf{U}_k(\mathbf{x}), \mathbf{f}_{k_{\text{Max}}}(\mathbf{x}))$;
8:   **if** $(|\mathbf{W}_\mathbf{x}| = 0)$ **then**
9:    $\widehat{I_k}(\mathbf{x}) \leftarrow \mathbf{M}_k(\mathbf{x})$
10:   **else**
11:    $i_S \leftarrow -1, \delta_{max} \leftarrow -1$
12:    **for** l in $\mathbf{W}_\mathbf{x}$ **do**
13:     $i \leftarrow \triangleright(l)$
14:     **while** $(i \leq \triangleleft(l))$ **do**
15:      $\delta_i \leftarrow HB\text{-}KDE(i, \lambda_k(\mathbf{x}), H_\mathbf{x}, \mathbf{L}_k(\mathbf{x}), \mathbf{U}_k(\mathbf{x}))$;
16:      **if** $(\delta_i > \delta_{max})$ **then**
17:       $\delta_{max} \leftarrow \delta_i, i_S \leftarrow i$
18:      **end if**
19:      $i \leftarrow i + \alpha$
20:     **end while**
21:    **end for**
22:    $\epsilon \leftarrow 1,\ i_0 \leftarrow (i_S - \frac{1}{2}\alpha),\ i_1 \leftarrow (i_S + \frac{1}{2}\alpha)$
23:    $\delta_0^\epsilon \leftarrow HB\text{-}KDE(i_0, \lambda_k(\mathbf{x}), H_\mathbf{x}, \mathbf{L}_k(\mathbf{x}), \mathbf{U}_k(\mathbf{x}))$;
24:    $\delta_1^\epsilon \leftarrow HB\text{-}KDE(i_1, \lambda_k(\mathbf{x}), H_\mathbf{x}, \mathbf{L}_k(\mathbf{x}), \mathbf{U}_k(\mathbf{x}))$;
25:    **repeat**
26:     **if** $(\delta_0^\epsilon > \delta_1^\epsilon)$ **then**
27:      $i_1 \leftarrow \frac{1}{2}(i_1 + i_0)$
28:      $\delta_1^\epsilon \leftarrow HB\text{-}KDE(i_1, \lambda_k(\mathbf{x}), H_\mathbf{x}, \mathbf{L}_k(\mathbf{x}), \mathbf{U}_k(\mathbf{x}))$;
29:     **else if** $(\delta_0^\epsilon < \delta_1^\epsilon)$ **then**
30:      $i_0 \leftarrow \frac{1}{2}(i_1 + i_0)$
31:      $\delta_0^\epsilon \leftarrow HB\text{-}KDE(i_0, \lambda_k(\mathbf{x}), H_\mathbf{x}, \mathbf{L}_k(\mathbf{x}), \mathbf{U}_k(\mathbf{x}))$;
32:     **else**
33:      $i_0 \leftarrow \frac{1}{2}(i_1 + i_0),\ i_1 \leftarrow i_0$
34:     **end if**
35:     $\epsilon \leftarrow \epsilon + 1$
36:    **until** $(((i_1 - i_0) > \beta) \wedge (\mathbf{max}(\delta_0^\epsilon - \delta_0^{\epsilon-1}, \delta_1^{\epsilon-1} - \delta_1^\epsilon) > \epsilon_0))$
37:    $\widehat{I_k}(\mathbf{x}) \leftarrow (i_0 \cdot \delta_0^\epsilon + i_1 \cdot \delta_1^\epsilon)/(\delta_0^\epsilon + \delta_1^\epsilon)$
38:   **end if**
39: **end if**
40: **end for**

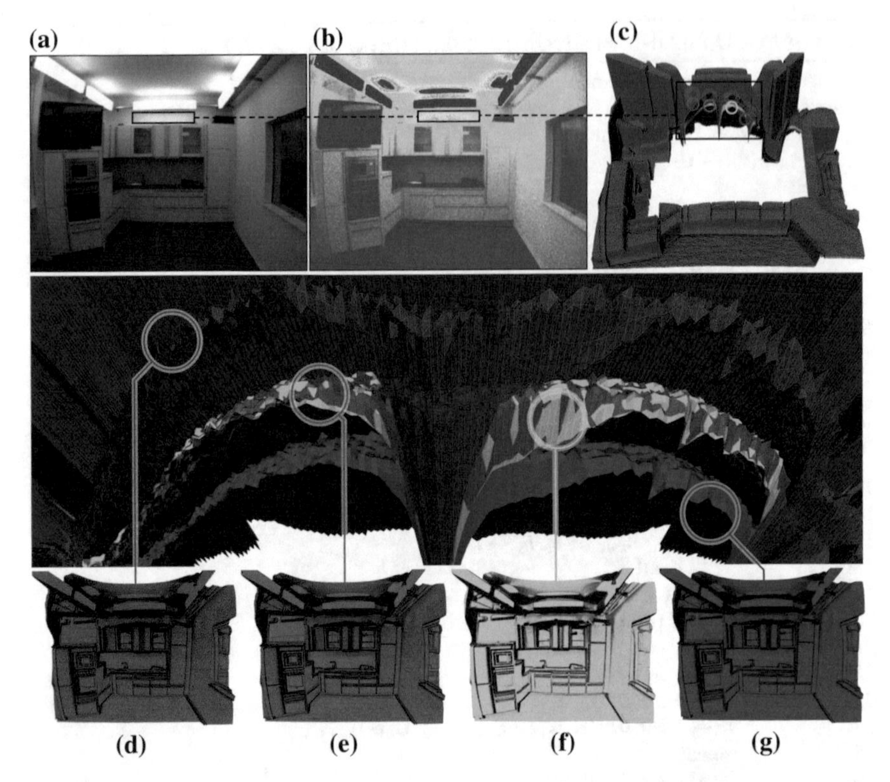

Fig. 4.7 Qualitative visualization of the optimal temporal fusion. **a** Semi-dynamic scene. **b** The image in pseudo-color denotes regions with wide deviation produced by the blinking lamp. **c** The irradiance descriptive statistics are displayed by means of a cross section. **d** The red surface shows the maximal intensity value that each pixel has sensed in the temporal scope (Eq. 4.3). **e** The magenta surface represents the mean value per pixel (Eq. 4.6). **f** The n-horizon fusion image $\check{I}_n$ (Sect. 4.5.9) in green (Eq. 4.9). The fusion and the mean intensity partially overlap (Sect. 4.5.4). The small deviation between mean and fusion values occurs due to the large sampling horizon $n = 400$. **g** The blue surface is the minimal sensed value (Eq. 4.4)

stages, the resulting n-horizon fusion image $\check{I}_n$ is regarded as the ideal reference for analyzing the convergence trade-off (in Sect. 4.5.10) and the sensor abnormality (in Sect. 4.5.11).

4.5.10 Convergence Trade-Off

In order to analyze the global quality improvement of the resulting fusion image relative to the amount of fused samples, a RMS discrepancy indicator between the n-horizon fusion image $\check{I}_n$ and the partial $(k < n)$ fusion image $\widehat{I}_k$ is expressed as

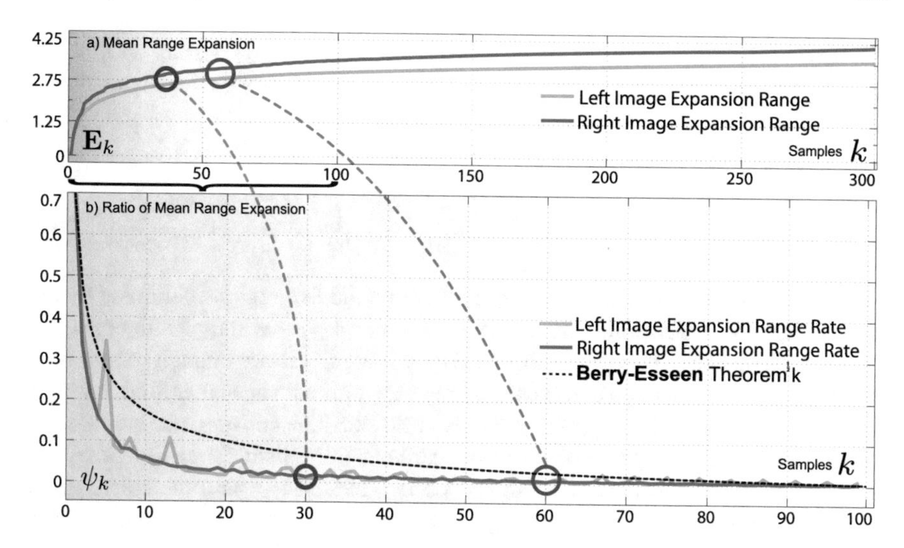

Fig. 4.8 The range expansion E_k and its rate Ψ_k provide an intuitive illustration of the sensor deviation convergence. The range expansion stability is reached at the time position where the curve shows a quasilinear tendency. It occurs approximately after $k > 60$ (blue circles), whereas the rate Ψ_k converges in a shorter (roughly) $k < 30$ temporal scope (magenta circle). Notice the subtle differences between the curves from the left and right images. The overall behavior is the same for all scenes, see support for this statement in Fig. 4.11. The black dotted line shows the convergence speed of the Berry-Esseen theorem [25]

$$\chi_k := \left(\frac{1}{wh} \left[\sum_{\mathbf{x}\in\Omega} \left(\widehat{I}_t(\mathbf{x}) - \breve{I}_n(\mathbf{x}) \right)^2 \right]_{t=1}^{k} \right)^{\frac{1}{2}}. \tag{4.18}$$

Figure 4.9 shows the discrepancy convergence of the density fusion processes. The convergence behavior χ_k is extensively evaluated (in Sect. 4.6.4) with diverse scenes and larger temporal scopes. The general convergence shape of the curves expose this tendency in a consistent manner. This is a substantial characterization for determining the trade-off between the amount of samples versus the quality of the resulting images. The *"optimal trade-off"* is estimated at the inflection point (c) in Fig. 4.9. Usually this point is located between the frames 24 to 30 depending on the image content. The capture and fusion of more images beyond point (c) implies longer sampling without significant improvements.

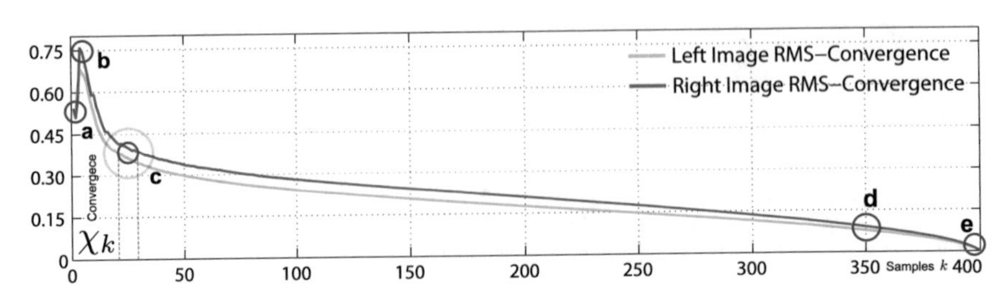

Fig. 4.9 The convergence behavior χ_k (Eq. 4.18). Note the five regions delimited by circles; **a** The initial deviation from χ_1 to χ_2. **b** After $4 \leq k \leq 6$ samples the maximal deviation is reached. This occurs because very few images are enough to introduce outliers but still not enough to robustly estimate the optimal visual manifold. **c** Within less than one sampling second ($15 \leq k \leq 25$ @ 30 FPS) the convergence slows down into a quasilinear behavior. **d** This quasilinear behavior remains for a long temporal scope. Finally, near to the end of the scope ($k > 350$, @ $n = 400$) the convergence behaves nonlinear. **e** Evidently, plenary convergence is reached at the sampling horizon, namely $k = n$

4.5.11 Distribution of Abnormal Behavior

The abnormality distribution describes the comprehensive spatio-temporal deviation of the intensity values. It provides an insight into the sensor anomalous distribution expressed by four curves. First, the intensity value $l \in \Theta$ has the upper abnormality

$$A_u(l) := \mathbf{max}\left[I_t(\mathbf{x}) - l \right]_{t=1}^{n}, \tag{4.19}$$

depicting the maximal intensity value found in the whole ($k = n$) temporal scope and spatial domain Ω corresponding to the intensity l in the discrete rounded version of the n-horizon fusion image $\breve{I}_n$. Likewise, the lower abnormality

$$A_l(l) := \mathbf{min}\left[I_t(\mathbf{x}) - l \right]_{t=1}^{n}, \tag{4.20}$$

the abnormality range $A_r(l) := A_U(l) - A_L(l)$ and the RMS-abnormality

$$A_\varsigma(l) := \left[\frac{1}{n} \sum_{t=1}^{n} (I_t(\mathbf{x}) - l)^2 \right]^{\frac{1}{2}}, \tag{4.21}$$

provide the complete deviation distribution. All these curves are subject to $l = \text{round}(\breve{I}_n(\mathbf{x})) : \forall \mathbf{x} \in \Omega$ (see Algorithm 5 and plots in Fig. 4.10).

Algorithm 5 Abnormality-Distribution, $AD(I_{1,...,}I_n,\breve{I}_n,A_u,A_l,A_r,A_\varsigma)$

Require: $n > 1$
1: $A_u \leftarrow [0]_{2^m},\ A_l \leftarrow [\infty]_{2^m},\ A_r \leftarrow [0]_{2^m},\ A_\varsigma \leftarrow [0]_{2^m},\ A_c \leftarrow [0]_{2^m}$
2: **for** $\mathbf{x}$ in Ω **do**
3: $\quad l \leftarrow \mathbf{round}(\breve{I}_n(\mathbf{x}))$
4: $\quad$ **for** $t = 1$ **to** n **do**
5: $\quad\quad A_u[l] \leftarrow \mathbf{max}(A_u[l], I_t(\mathbf{x})),\ A_l[l] \leftarrow \mathbf{min}(A_l[l], I_t(\mathbf{x}))$
6: $\quad\quad A_\varsigma[l] \leftarrow A_\varsigma[l] + (I_t(\mathbf{x}) - l)^2,\ A_c[l] \leftarrow A_c[l] + 1$
7: $\quad$ **end for**
8: **end for**
9: **for** $l = 0$ **to** $(2^m - 1)$ **do**
10: $\quad$ **if** $(A_c[l] > 0)$ **then**
11: $\quad\quad A_r[l] \leftarrow A_u[l] - A_l[l],\ A_\varsigma[l] \leftarrow \sqrt{A_\varsigma[l]/A_c[l]}$
12: $\quad$ **end if**
13: **end for**

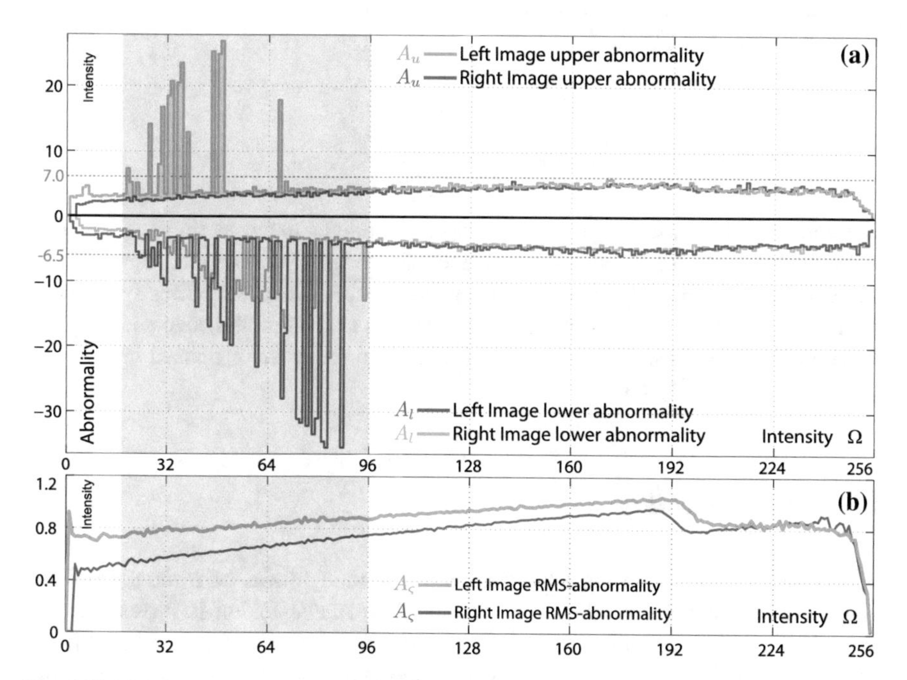

Fig. 4.10 The abnormality plots show the relation between noise distribution and intensity values. **a** The abnormality extrema (from Fig. 4.2) show the distributions of both cameras. **b** Shows the RMS-abnormality A_ς. In the gray region, the prominent outliers on the upper A_u and lower A_l abnormality produce detrimental effects for feature extraction and image segmentation. Notice that this outliers are not representative in the RMS-abnormality. This supports the proposed approach for the pixel-wise fusion

4.6 Experimental Evaluation: Image Fusion

The generality of the convergence behavior is experimentally analyzed (in Sect. 4.6.1) in order to support the strategy (discussed in Sect. 4.5.10). Afterwards, in order to simultaneously support the contribution of this method and quantitatively evaluate the improvement effects, two important feature extractions are used to evaluate the improvement in terms of stability (see Sects. 4.6.2 and 4.6.3).

4.6.1 Convergence Behavior

The mobile kitchen (from Fig. 3.23) was used to evaluate the convergence of the fusion method with 32 different viewpoints and states (Sect. 3.43) including opened and closed doors, drawers and electric appliances (see Fig. 4.12). The process described in Sect. 4.5.10 was applied for each of these scenes. The quantitative results are shown in Fig. 4.11.

4.6.2 Segmentation Stability

The noise artifacts within the directly captured images (such as those in the gray region in Fig. 4.10) have negative side effects on image segmentation. These effects connect discontinuous segments and vice versa. Hence, when using directly captured images, the segmentation is not consistent in terms of the amount, size and shape of the resulting segments. These issues can be overcome by the proposed method for image fusion (see Fig. 4.13).

4.6.3 Edge Stability

Due to the importance of the edge cue for visual recognition, an evaluation of the stability improvements produced by the multiple image fusion is presented (see Fig. 4.14). The importance of the stability and precision of the image intensity is directly reflected during saliency estimation and non-maximal suppression for edge extraction. This occurs because independently of the edge response function, the subtle differences produced by noise alterations can affect or vanish the edges from the images.

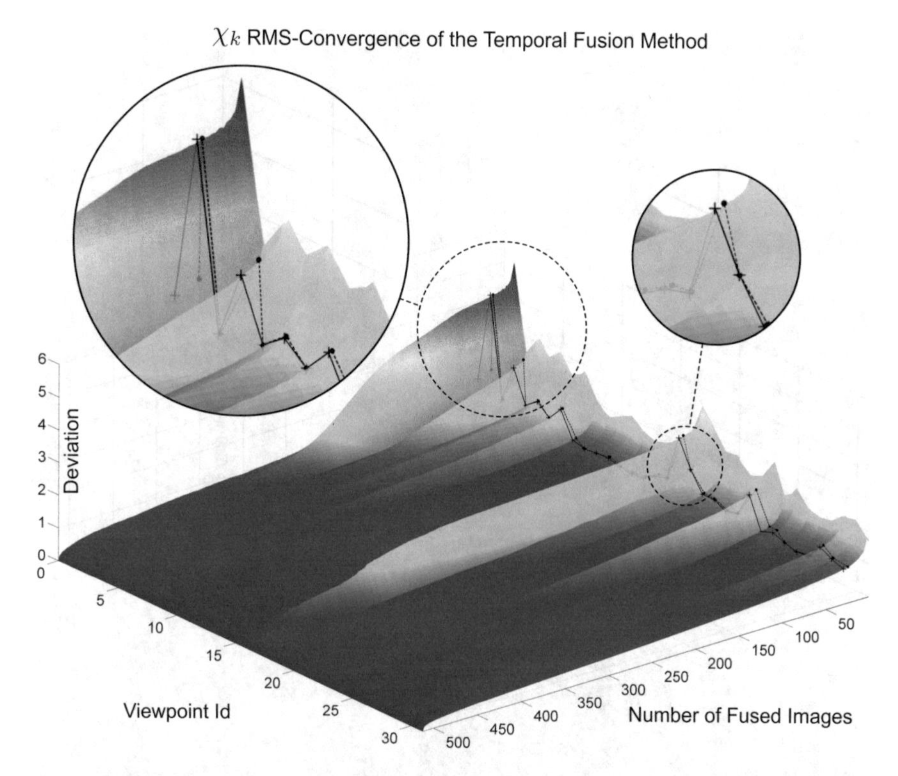

Fig. 4.11 The convergence behavior χ_k (Eq. 4.18) with 32 different viewpoints (from the setup in Fig. 4.12). There are viewpoints which slightly differ from the mean behavior. However, even in at these viewpoints the inflection points have the same shape in terms of sequence and relative convergence. The convergence behavior points (marked with black circles) are no further than frames 30 (marked with black crosses) for all viewpoints. This supports the selected convergence criterion (in Sect. 4.5.10)

4.6.4 Convergence Trade-Off Versus Integration Time

The proposed fusion method is capable of coping with various sensing problems including the non-uniformity of pixels, on- and off-chip noise as well as diverse environmental issues. However, the iso-exposure nature of the method limits the dynamic range of the images. This problem is managed in the next complementary method (see Sect. 4.7) by integrating differently exposed images. This requires precise control of the integration time while capturing image sets. In order to keep the method practicable for online applications, the amount of images with different exposures is managed by the following analysis.

Based on the convergence trade-off indicator (Eq. 4.18), it is possible to determine the fusion deviation level according to the amount of fused images. This analysis includes both the amount of images to fuse and their exposure time. This provides additional means to cope with the *dark-current* (thermal drifting of the sensor) and

Fig. 4.12 Evaluation set with 32 viewpoints for convergence strategy validation. At each viewpoint, 512 images were taken to obtain n-horizon fusion images $\check{I}_n$ and corresponding χ_k convergence curves (Eq. 4.18)

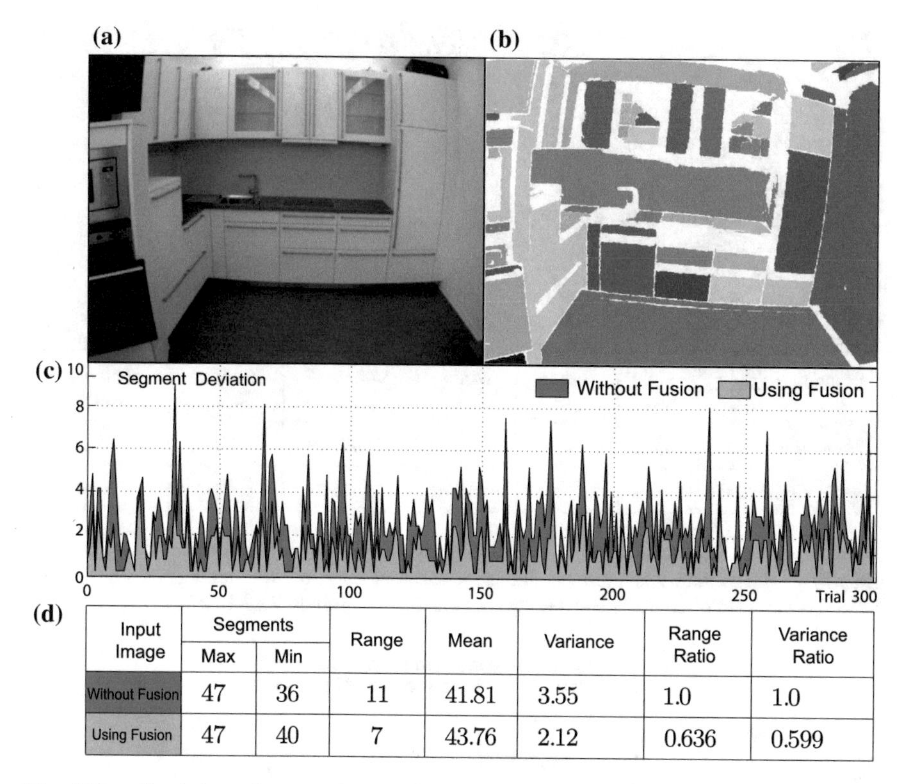

Input Image	Segments		Range	Mean	Variance	Range Ratio	Variance Ratio
	Max	Min					
Without Fusion	47	36	11	41.81	3.55	1.0	1.0
Using Fusion	47	40	7	43.76	2.12	0.636	0.599

Fig. 4.13 **a** Semi-dynamic scene. **b** A trial resulting from the segmentation based on adaptive region growing (similar method to [29]). **c** The absolute variation of segments relative to the mean of all trials. **d** Results show a **36.36%** reduction in the variation range of segments and a **40.05%** less variance using images resulting from the proposed fusion method

shot-noise effects (signal-to-noise quantization effects noticeable at short integration times [31]). The results of the analysis (see Fig. 4.15) are used to select the amount of images to be fused depending on the target exposure. This ensures the consistency of the acquired visual manifold (see Sect. 4.7.6).

4.6.5 *Performance of Temporal Image Fusion*

The computation of the fusion method begins with the sensing integration (Algorithm 1). Afterwards, the optimal intensity estimation requires to extract the pixel-wise optimization interval set (Algorithm 2) in order to (α-sparsely) approximate the PDF by means of the histogram based KDE (Algorithm 3). Finally, a β-refinement is done by the optimal fusion (see Algorithm 4) with execution times shown in the Sect. 4.6.5. In summary, the fusion process determines the optimal *"noiseless"*

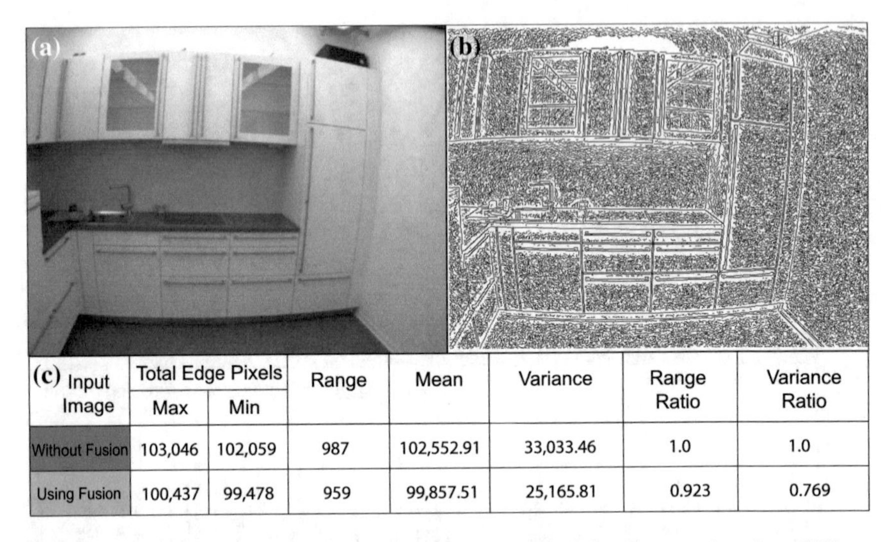

(c) Input Image	Total Edge Pixels		Range	Mean	Variance	Range Ratio	Variance Ratio
	Max	Min					
Without Fusion	103,046	102,059	987	102,552.91	33,033.46	1.0	1.0
Using Fusion	100,437	99,478	959	99,857.51	25,165.81	0.923	0.769

Fig. 4.14 **a** Semi-dynamic scene. **b** A resulting trial of the edge detector (based on [30], see Chap. 5). **c** Results show **23.81%** reduction in the variation range of the amount of edge pixels and **7.7%** less variance when using fused images attained by the proposed method

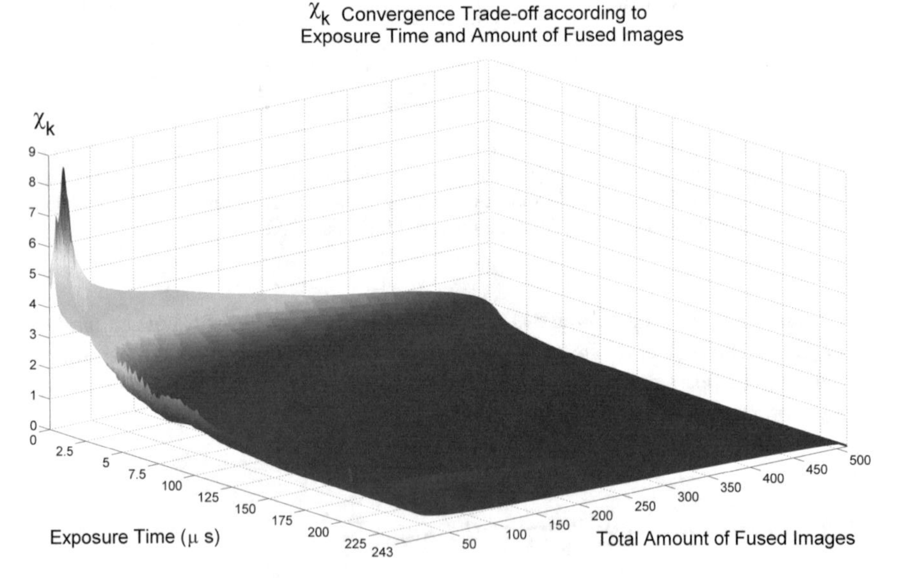

Fig. 4.15 The convergence trade-off (Eq. 4.18) using 180 different exposure times with 512 images per fusion. This plot shows the expected behavior resulting from the dark-current and shot-noise effects. In order to achieve high quality fusion images, when capturing short exposed images, more samples are required compared to long exposed images. The particular assertion of the amount of images is presented (in Sect. 4.7.5) and is coordinated by the camera radiometric calibration and distribution of the exposure set

Table 4.1 Running time performance of the proposed fusion method. These results were obtained with a CPU Intel(R) Core(TM)2 Quad @ 2.33 GHz in a non-optimized single thread implementation

Phase (ms)	Maximal	Minimal	Mean	Deviation
Iso-Exposed capturing	2.25	2.06	2.11	0.04
Iso-Exposed fusion (30 Frames)	891.89	712.05	762.94	55.92
Reset data structure	16.60	12.72	13.31	0.44

image. At the same time, the image fusion obtains continuous range intensity values. This enables more precise feature extraction computations. However, due to the floating point nature of the intensity, the process performs slower than its integer version when using standard CPUs. This is a critical drawback of the methods. This is a technical limitation manageable using DSPs or multiple core processors (see evidence supporting this statement in the experimental performance using GPU in [32]) (Table 4.1).

4.7 Dynamic Range

Humanoid robots have to be able to visually explore and recognize objects in their environment for self-localization, scene understanding, grasping and manipulation. In these active fields much research has been done. Consequently, considerable results have been recently achieved and several humanoid robots expose elaborated visual recognition capabilities [33–38]. However, all these approaches assume restricted light conditions. These assumptions are attained either by controlling the light or restricting the 6D-pose of the robot during recognition.

Due to these limitations, the applicability of humanoid robots is strongly reduced in the presence of everyday wide intra-scene radiance range or dynamically varying illumination (for instance, the back lighting of a window, the interior of an oven, the complex structures inside a dishwasher or the metallic surface of a microwave, see Fig. 4.16).

Avoiding the ubiquitous high-contrast image content in everyday applications is not plausible. Neither is it possible to mend the dynamic range issues through the camera auto-exposure in order to allow recognition approaches to implicitly hold the illumination assumptions. These issues arise because the adaptively selected exposure attempts to preserve the majority of the image content at the cost of losing the extremal regions. This condition generates severe rate distortion quantization effects which substantially diminish the image quality. As a result, local over- or/and under-exposure occurs. Additionally, even if the objects are completely captured

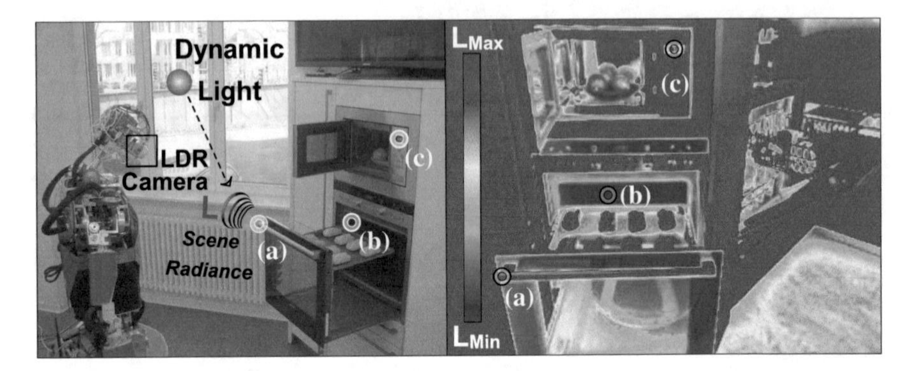

Fig. 4.16 The ubiquitous wide intra-scene radiance range. In made-for-humans environments, humanoid robots are confronted with high-contrast scenes, dynamically varying illuminations and complex materials. On the left, the humanoid robot ARMAR-IIIa [4] encounters a wide range illumination scene in a household environment. On the right, the high dynamic range (HDR) image captured with the low dynamic range (LDR) camera of the robot using the radiometric calibration and exposure control presented in Sect. 4.7.1. The pseudo-color representation allows a better visualization of the **a** bright, **b** dark and **c** complex regions which require precise sensing mechanisms

within *unclipped* image regions (where the discrete values are neither *white-* nor *black*-clipped), the resulting intensity values are still rarely a true measurement of the relative radiance of the scene.

Under such non-injective circumstances, a single discrete intensity value comprises a wide range of radiances. This many-to-one mapping makes it unfeasible to robustly and accurately extract structured visual cues (geometric saliencies such as edges or corners) even when using computationally expensive approaches such as automatic space-scale selection [39]. Without these essential visual cues, the recognition with pose estimation becomes notably complicated or intractable. In Sect. 4.7.1, these limitations were overcome by an HDR image acquisition which robustly and accurately manages the intra-scene radiance range (see Fig. 4.4).

4.7.1 HDR Image Acquisition

Until now, HDR cameras are not widespread, especially in applications with several tight constraints, for instance, high frame rate, light weight, reduced space, low power consumption and compliance with saccadic movements (see [40]). These are a few restrictions in the eyes of most humanoid robots, for example, the humanoid robots ARMAR-IIIa,b or the iCub robot [41] (see their detailed compositions in [5, 42]). Due to these restrictions, an appropriate mechanism is introduced to capture HDR images by suitably employing the low dynamic range (LDR) cameras of these humanoid robots. Conceptually, the process of attaining HDR images is rather simple. During the image synthesis, the short-exposed images sample the high radiance regions of the scene, inversely, the long-exposed images sample the low radiance regions. The integration of this information in an optimal trade-off is done by means of the radiometric response function (see [43]) of the whole optical system. Hence, the mechanism for attaining HDR images consists of the following elements: First, the nonlinear transformation mapping the scene radiance to the discrete intensity values in the image is attained (the radiometric calibration). Second, the robust exposure control by analysis of the integration granularity of the camera is performed. The coordinated integration of these elements captures the intra-scene radiance range by fusing a minimal collection of differently exposed images. The strategy to efficiently synthesize high quality HDR images is to analyze the radiometric calibration and granularity of the integration times of the camera(s) in order to determine the optimal exposure selection. This reduces the total amount of synthesized images and ensures the HDR image quality.

4.7.2 Radiometric Calibration: Related Work

During the 90s, various important contributions were made on high-dynamic-range scene capturing using digital cameras. Computational photography methods focused on the radiometric calibration and dynamic range expansion process. In particular, the work of P. Mann et al. [23] is an important contribution to this field. It set the foundation for response function estimation based solely on images. The key idea of that approach was the modeling of the camera response function without using controlled light sources or other complex luminance devices. Further, in [23] the authors presented exposure *bracketing*[4] as the control element during calibration. Despite the influential ideas of this work, the dependency on particular image content forbids the systematic application of the proposed algorithm in general scenarios. Later, the camera response function (from now on called radiometric calibration) was obtained without assumptions about the image content. This method was presented in the work of P. Debevec et al. [44]. The method was designed for digital cameras with precise controllable integration timing. The key idea of the approach was to analyze the progression of the intensity values as a monotonic function depending on the integration time. This fundamental idea is found in almost all following calibration methods (see [45, 46]). Recently, novel methods have appeared (for instance [47]) which only use a single image to attain the radiometric calibration. Nevertheless, they assume certain image content (as in [23]) such as colored edges or fixed patterns. Thus, their application in humanoid robots remains limited. The general mathematical approach of the method from P. Debevec et al. established a fundamental research reference in the field. Because this method has no image content dependency, it is possible to use any camera supporting the DCAM/IIDC specification [48] with precise integration timing. Despite being suitable for humanoid robots, the Debevec method cannot be directly applied for structural feature extraction (see Fig. 4.19). It occurs because of

[4] Capturing several images of the scene while varying a camera parameter.

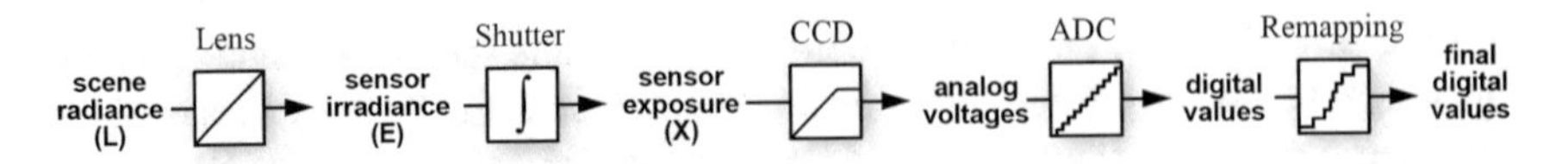

Fig. 4.17 Image formation pipeline from [44] [©1997 -ACM]

two critical issues, namely λ-*smoothing* and *reciprocity consistency*. In order simultaneously explain these issues and clearly separate the original work in [44] from the improvements realized in this work (see also [49]), a sequential presentation of the Debevec method is given (in Sect. 4.7.3) followed by the criticism and proposed improvements (in Sect. 4.7.4).

4.7.3 Debevec Radiometric Calibration

The intensity values provided by a digital camera are ideally subject to a principle called *reciprocity*. In order to explain this principle the following definitions are required: The sensor exposure X is the collected energy per surface unit in Jm^{-2} (see Fig. 4.17). This collected energy is ambiguously called exposure. It is the product of sensor *irradiance*[5] E and integration time Δt

$$X = E\Delta t. \tag{4.22}$$

The reciprocity principle states: The total collected energy remains constant $X' = X$ if the sensor irradiance is scaled $E' = \alpha E$ as has long as the integration time is inversely scaled $\Delta t' = \frac{1}{\alpha}\Delta t$. The reciprocity principle does not entirely hold in physical systems (see Hurter-Driffield curve [50]) particularly at the upper and lower bounds of the sensor irradiance. Figure 4.17 shows that once the sensor irradiance has been integrated, the electronic components of the camera convert the analog signal to its discrete and quantized version. This composed process is followed by a nonlinear mapping (usually gamma compression) for the high contrast content to be properly represented within the dynamic limits of the $m = 8$ bits image. This nonlinear mapping rescales the signals to better fit human perceptual metrics. The whole process is called the *sensor-mapping function* and it is denoted as $\mathcal{F}$. The pixel-wise sensor-mapping function can be expressed as

$$I_{\mathbf{x}}^{i} = \mathcal{F}(E_{\mathbf{x}}\Delta t_i), \tag{4.23}$$

where the irradiance at the pixel location $\mathbf{x} \in \Omega$ is denoted as $E_{\mathbf{x}}$ (Eq. 4.1). The exposure time Δt_i is controllable and the intensity value $I_{\mathbf{x}}^{i} \in \Theta$ (Eq. 4.2) is provided by the camera interface [51]. Because the mapping function $\mathcal{F}$ is monotonic, it is possible to find its inverse function such as

[5] The scene radiance modulated by the optical response function (see Fig. 4.17).

$$E_{\mathbf{x}}\Delta t_i = \mathcal{F}^{-1}(I^i_{\mathbf{x}}). \tag{4.24}$$

The ideal estimation of the sensor irradiance from the pixel $\mathbf{x}$ is $E_{\mathbf{x}} = \mathcal{F}^{-1}(I^i_{\mathbf{x}})/\Delta t_i$. In [44], considerations were taken for modeling $\mathcal{F}^{-1}$:

- **Inverse function as subspace**: Because the codomain of $\mathcal{F}$ is the finite intensity set Θ, the estimation of $\mathcal{F}^{-1}$ considers the function as a vector $\mathcal{F}^{-1} \in R^{2^m}$ in least squares formulation.
- **Noise tolerant formulation**: The sensor noise is considered during the estimation process. This is coherently formulated by kernel weighting also in least squares error sense.

The formulation obtaining $\mathcal{F}^{-1}$ was proposed by Debevec et al. as follows: First, taking the natural logarithm of Eq. 4.24 and renaming[6] $g = \ln \mathcal{F}^{-1}$ the resulting expression is

$$g(I^i_{\mathbf{x}}) = \ln E_{\mathbf{x}} + \ln \Delta t_i. \tag{4.25}$$

Equation 4.25 is tractable when considering both: (i) There are only 2^m possible values for the function g (recall m denotes usually 8 bits per pixel Eq. 4.2). (ii) The "*stops*" term $\ln \Delta t_i$ is known for all integration times. Moreover, taking ρ images with different integration times and the selecting a subset of pixel $S \subset \Omega$ (see Eq. 4.1), the minimization quadratic objective function for obtaining g is expressed

$$O = \sum_{i}^{\rho} \sum_{\mathbf{x} \in S} \left[g(I^i_{\mathbf{x}}) - \ln E_{\mathbf{x}} - \ln \Delta t_i \right]^2. \tag{4.26}$$

Additionally, in order to ensure the evenness of the estimated function g, Debevec et al. added a λ-smoothing restriction term as

$$O = \sum_{i}^{\rho} \sum_{\mathbf{x} \in S} \left[g(I^i_{\mathbf{x}}) - \ln E_{\mathbf{x}} - \ln \Delta t_i \right]^2 + \lambda \sum_{u=1}^{2^m-1} \left(g(u-1) - 2g(u) + g(u) \right)^2. \tag{4.27}$$

Furthermore, because the reciprocity is not held at the intensity extrema, the contribution of the values are weighted according to their intensity with less ponderation at the extrema. This was done by a symmetric kernel $w : \Omega \mapsto \mathbb{R}$ with maximal central value $w(\frac{1}{2}(2^m - 1)) = 1$. In [44], the authors propose a triangular kernel. However, it has been discussed in [46], the use of a Gaussian kernel (or other continuous kernel with less support at the extrema) enables better radiometric calibration results. This kernel weighting allows the coupled and smooth estimation of the function g as

[6]The aim of the alias function g is to reduce the notation.

$$O = \sum_{i}^{\rho} \sum_{\mathbf{x} \in S} \left\{ w(I_{\mathbf{x}}^{i}) \left[g(I_{\mathbf{x}}^{i}) - \ln E_{\mathbf{x}} - \ln \Delta t_i \right] \right\}^2 + \lambda \sum_{u=1}^{2^m-1} \left(w(u) \left[g(u-1) - 2g(u) + g(u) \right] \right)^2 . \tag{4.28}$$

The estimation of the radiometric calibration (as the weighted square minimization problem in Eq. 4.28) is done by a SVD (Singular Value Decomposition) least squares formulation (see appendix in [44]). Particularly, the computation of the overdetermined system requires that the amount of pixels and different exposures to be at least $\rho(|S| - 1) \geq 2^m$. Commonly (when $m = 8$), the system is properly overdetermined by capturing 16 different exposures with 32 strategically distributed pixels.

4.7.4 *Improved Radiometric Calibration*

The systematic application of the Debevec radiometric calibration method shows critical issues:

- **λ-Smoothing**: First, the smoothing factor λ changes the resulting curves without implicit nor explicit selection criterion (see Fig. 4.18).
- ***S*-Sampling-set**: Second, another issue is the selection criterion for pixels in the sampling set S (see Eq. 4.26). Still, these two drawbacks can be iteratively and experimentally solved.
- **Reciprocity consistency**: Third, there is a more complex intrinsic issue. The resulting g curves present noisy artifacts that cannot be removed by varying the calibration parameters (λ, S, ρ) of the Debevec formulation.

These noisy curves produce detrimental effects near to the upper and lower intensity ends. In Fig. 4.19, these calibration artifacts appear because the reciprocity between exposure and intensities is not held. This is the jointed product of numerical (least squares fitting) effects and intrinsic physical sensor behavior. The results are noisy calibrations propagating into salient artifacts in the HDR image (see Fig. 4.19). In order to anticipate all these issues, the optimal selection of the smoothing factor and a new improvement extension of the radiometric calibration were simultaneously realized as follows.

Continuous Reciprocity-Consistent Calibration Model

In order to unveil the radiometric artifacts (see Fig. 4.19), the calibration function is transformed by mapping the λ-*optimal smooth*[7] function g_λ from its *homomorphic*[8]

---

[7]The optimality smoothing criterion is attained in lines 6–14 of Algorithm 6.

[8]Homomorphic filtering are methods for signal processing based on nonlinear mappings to target domains in which linear filtering is applied followed by the corresponding back mapping to initial domain (see [52]).

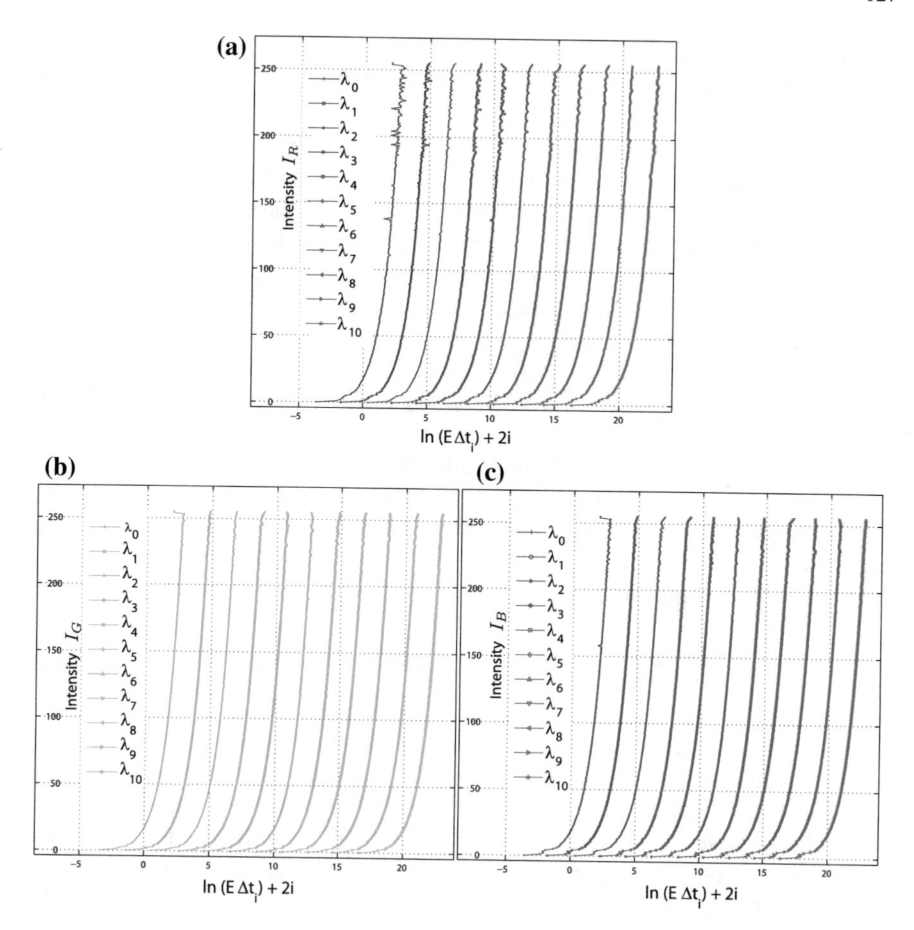

Fig. 4.18 The Debevec radiometric camera calibration. **a** The λ-smoothing curves of the red-channel show that this is the most noisy channel. **b** The λ-smoothing curves of the green-channel. This channel has the higher signal-to-noise ratio of all three channels. **c** The blue-channel λ-smoothing curves. For each λ_i plot, the horizontal axes were right shifted $2i$ units to conveniently display the comparative effects of the λ-smoothing factor

domain [53] to the lineal domain through the exponential transformation

$$g_{\lambda}^{*}(u) := \mathbf{exp}_{u}\left[g_{\lambda}(u)\right], \tag{4.29}$$

where the auxiliar variable $0 \leq u \leq (2^m - 1)$ covers the intensity set Θ. The resulting functions $g^*_\lambda(u)$ of each color channel[9] clearly illustrates the noise at the upper region of the intensity set (see Fig. 4.19a).

Additionally, the calibration curve g_λ has a *"limited discrete"* domain. This is a deficiency using images with real intensity values resulting from the fusion method (see Sect. 4.5). Based on these observations, the calibration curve g^*_λ can be improved by finding a continuous function g^L_λ which tightly fits the tendency of g^*_λ. This fitting process considers the intensity regions according to their reciprocity confidence by gradually weighting their contributions. This leads to a kernel weighted regression for obtaining a continuous model

$$g^L_\lambda : \mathbb{R} \mapsto \mathbb{R} \; ; \; g^L_\lambda(u) \cong g^*_\lambda(u), \tag{4.30}$$

see detailed presentation in Algorithm 6 and results in Fig. 4.20.

In contrast to [23, 44, 46], the improved radiometric calibration function properly holds the reciprocity assumption even at the extremal intensity (see results in Fig. 4.21).

The Gaussian kernel defines as

$$\mathcal{N}\left(\mu := \frac{1}{2}(2^m - 1),\ \sigma := \frac{1}{3}(2^m - 1)\right) \tag{4.31}$$

allows proper integration of the information within the smooth region of the curve g^*_λ while gradually disregarding the broken reciprocity regions. In this manner, the calibration deviations and detrimental artifacts were soundly removed.

In addition, a novel feature of the calibration model g^L_λ is to allow the estimation of the sensor response function at continuous intensities resulting from the temporal fusion of k iso-exposed images as

$$g^M_\lambda(\widehat{I^i_k}) := \ln\left(\widehat{I^i_k} \cdot g^L_\lambda[1,1] + g^L_\lambda[1,2]\right), \tag{4.32}$$

where $\widehat{I^i_k}$ is the i-exposed temporal fusion (Eq. 4.9) and g^M_λ is the proposed radiometric calibration model in the standard homomorphic domain. The improved radiometric calibration is unique per color channel and (due to the logarithm) is restrictively defined within the interval

$$g^M_\lambda : \left(\mathbf{max}(-g^L_\lambda[1,2]/g^L_\lambda[1,1], 1), (2^m - 1)\right] \in \mathbb{R}^+ \mapsto \mathbb{R}. \tag{4.33}$$

In summary, the radiometric calibration model g^M_λ proposed in this section is an improvement and extension of the method [44]. This model calibration does not only addressed the parameter selection in optimal manner. It also provides a continuous

[9]Due to their different spectral responses, a separated calibration per color channel (R,G,B) is done when using Bayer pattern sensors (see [44]).

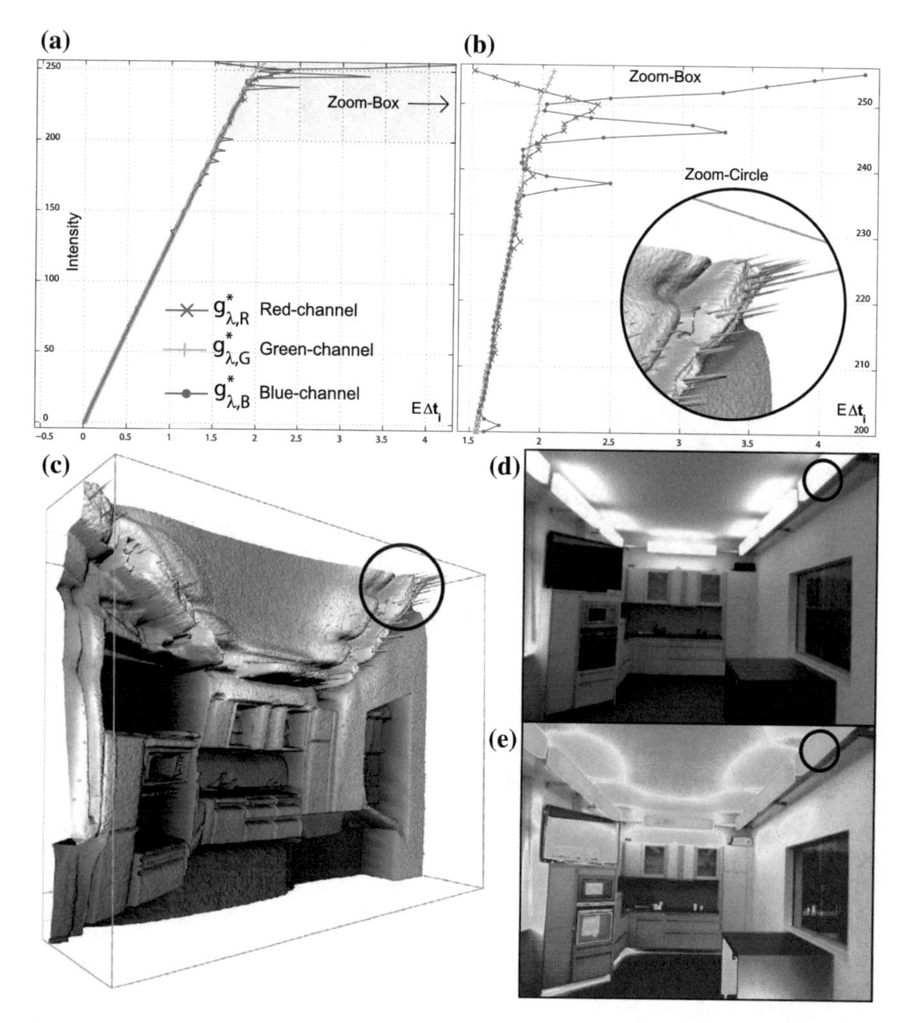

Fig. 4.19 The exponential domain of the radiometric calibration allows the detection of detrimental artifacts. **a** The exponential calibration curve g^*_λ per color channel. Within the light gray region, the Debevec calibration produces detrimental peaks. **b** The zoom-box and the zoom-circle show these negative effects. **c** Prominent artifacts in the HDR image are not detected by the human perception in the radiance map. However, these artifacts severely impact the structural feature extraction for object recognition. **d** Everyday scene in made-for-humans environment. **e** Pseudo-color HDR image

Algorithm 6 Radiometric-Calibration-Model, $RCM(S,\rho,\epsilon,\delta_\lambda)$

Require: $\rho(|S|-1) \geq (2^m)$ **and** $0 < \delta_\lambda \leq \frac{1}{10}$
1: $P \leftarrow [0]_{2^m \mathrm{x} 2^m},\ X \leftarrow [0]_{2^m \mathrm{x} 2},\ \lambda_s \leftarrow 0,\ \lambda_{\min} \leftarrow 0,\ E_{\min} \leftarrow \infty$
2: **for** $u = 0$ **to** $(2^m - 1)$ **do**
3: $\quad X[u] \leftarrow [u, 1],\ P[u,u] = \mathbf{exp}\,[\frac{-3(u-2^{m-1})^2}{2^{(m+1)}-1}]$
4: **end for**
5: $K \leftarrow [X^T P X]^{-1}[X^T P]$
6: **while** $\lambda_s \leq 1$ **do**
7: $\quad g^*_{\lambda_s} \leftarrow \mathbf{exp}_u\,[Debevec(\lambda_s,S,\rho)],\ M \leftarrow K g^*_{\lambda_s},\ \lambda \leftarrow \lambda + \delta_\lambda,\ E \leftarrow 0$
8: $\quad$ **for** $u = 0$ **to** $(2^m - 1)$ **do**
9: $\qquad \delta \leftarrow (uM[1,1] + M[1,2]) - g^*_{\lambda_s}(u),\ E \leftarrow E + \delta^2$
10: $\quad$ **end for**
11: $\quad$ **if** $(E < E_{\min})$ **then**
12: $\qquad E_{\min} \leftarrow E,\ \lambda_{\min} \leftarrow \lambda_s$
13: $\quad$ **end if**
14: **end while**
15: $\lambda_a \leftarrow \lambda_{\min} - \frac{\delta_\lambda}{2},\ g^*_{\lambda_a} \leftarrow \mathbf{exp}_u\,[Debevec(\lambda_a,S,\rho)],\ M_a \leftarrow K g^*_{\lambda_a},\ E_a \leftarrow 0$
16: $\lambda_b \leftarrow \lambda_{\min} + \frac{\delta_\lambda}{2},\ g^*_{\lambda_b} \leftarrow \mathbf{exp}_u\,[Debevec(\lambda_b,S,\rho)],\ M_b \leftarrow K g^*_{\lambda_b},\ E_b \leftarrow 0$
17: **for** $u = 0$ **to** $(2^m - 1)$ **do**
18: $\quad \delta_a \leftarrow (uM_a[1,1] + M_a[1,2]) - g^*_{\lambda_a}(u),\ E_a \leftarrow E_a + \delta_a^2$
19: $\quad \delta_b \leftarrow (uM_b[1,1] + M_b[1,2]) - g^*_{\lambda_b}(u),\ E_b \leftarrow E_b + \delta_b^2$
20: **end for**
21: **repeat**
22: $\quad$ **if** $(E_a < E_b)$ **then**
23: $\qquad E_b \leftarrow 0,\ \lambda_b \leftarrow \frac{1}{2}(\lambda_a + \lambda_b)$
24: $\qquad g^*_{\lambda_b} \leftarrow \mathbf{exp}_u\,[Debevec(\lambda_b,S,\rho)],\ M_b \leftarrow K g^*_{\lambda_b}$
25: $\qquad$ **for** $E_b \leftarrow 0,\ u = 0$ **to** $(2^m - 1)$ **do**
26: $\qquad\quad \delta_b \leftarrow (uM_b[1,1] + M_b[1,2]) - g^*_{\lambda_b}(u),\ E_b \leftarrow E_b + \delta_b^2$
27: $\qquad$ **end for**
28: $\quad$ **else**
29: $\qquad E_a \leftarrow 0,\ \lambda_a \leftarrow \frac{1}{2}(\lambda_a + \lambda_b)$
30: $\qquad g^*_{\lambda_a} \leftarrow \mathbf{exp}_u\,[Debevec(\lambda_a,S,\rho)],\ M_a \leftarrow K g^*_{\lambda_a}$
31: $\qquad$ **for** $E_a \leftarrow 0,\ u = 0$ **to** $(2^m - 1)$ **do**
32: $\qquad\quad \delta_a \leftarrow (uM_a[1,1] + M_a[1,2]) - g^*_{\lambda_a}(u),\ E_a \leftarrow E_a + \delta_a^2$
33: $\qquad$ **end for**
34: $\quad$ **end if**
35: **until** $((\lambda_b - \lambda_a) > \epsilon)$
36: $g^*_\lambda \leftarrow \mathbf{exp}_u\,[Debevec(\frac{E_b\lambda_a + E_a\lambda_b}{\lambda_a + \lambda_b},S,\rho)]$
37: **return** $g^L_\lambda \leftarrow K g^*_\lambda$

and computational efficient radiometric function which enables the integration of the whole reachable radiance without producing detrimental artifacts. Because the reciprocity is not physically held at sensor level, the kernel weighting used during regression (Algorithm 6) must be applied during the HDR synthesis, namely the integration of the differently exposed images (see Sect. 4.7.5).

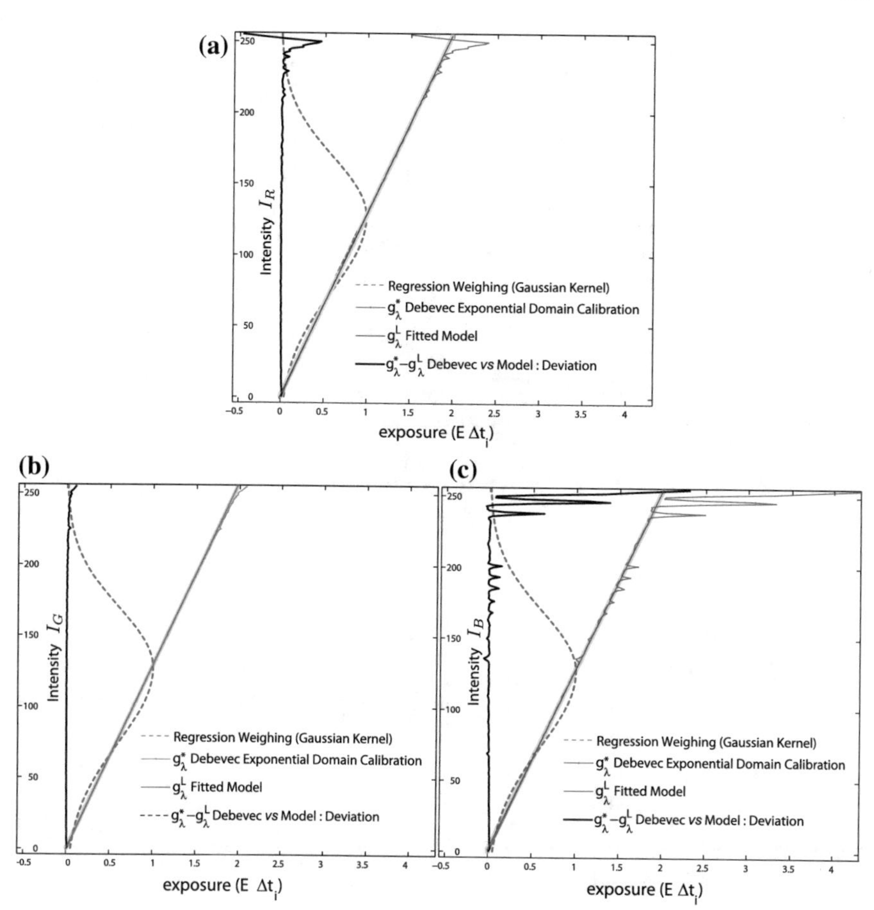

Fig. 4.20 The model calibration g^L_λ is accomplished by kernel weighted regression in the exponential domain. The Gaussian kernel $\mathcal{N}$ (Eq. 4.31) ponderates the λ-optimal g^*_λ calibration in order to estimate the model g^L_λ. The model calibrations are: **a** $g^L_{\lambda,R}(I_r) = \ln(0.007961 * I_r) - 0.001419$ @ $\lambda_R = 0.94726$. **b** $g^L_{\lambda,G}(I_g) = \ln(0.007602 * I_g) + 0.003388$ @ $\lambda_G = 0.95421$. **c** $g^L_{\lambda,B}(I_b) = \ln(0.007798 * I_b) + 0.001609$ @ $\lambda_B = 0.96014$. These results were obtained using 16 images with 273 pixels per image. The pixel locations were selected by the maximal intensity variance with a 32 pixel radius dominance

4.7.5 HDR Image Synthesis

The visual manifold acquisition is realized as follows: (i) For a initial integration time Δt_1 a set of k_1 images is captured in order to estimate the temporal fusion image $\widehat{I}^1_k$ (see Eq. 4.9). (ii) Next, the following integration time Δt_2 is set and the next set of images with cardinality k_2 is taken in order to obtain the fusion image $\widehat{I}^2_k$. (iii) In a similar manner as in [44], the collection of n temporal fused images $\{\widehat{I}^1_k, ..., \widehat{I}^n_k\}$ and the radiometric calibration model g^M_λ are used to synthesize the HDR image E as

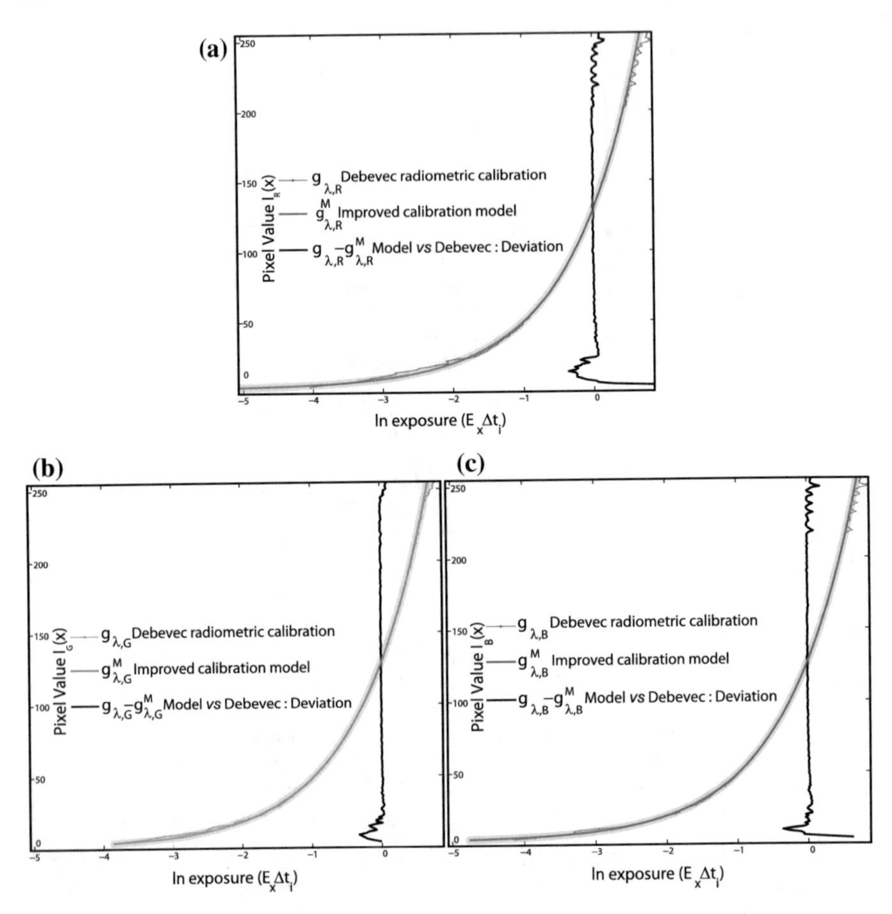

Fig. 4.21 The plots of the continuous reciprocity-consistent calibration model g_λ^M and their comparison with the calibrations obtained with the method in [44]. The deviation plots show the improvements, namely the removed detrimental effects

$$E_{\mathbf{x}} = \mathbf{exp}\left[\frac{\sum_{i=1}^{n} \mathcal{N}\left(\widehat{I}_k^i(\mathbf{x})\right) \ln\left(\widehat{I}_k^i(\mathbf{x}) \cdot \frac{g_\lambda^L[1,1]}{\Delta t_i} + \frac{g_\lambda^L[1,2]}{\Delta t_i}\right)}{\sum_{i=1}^{n} \mathcal{N}\left(\widehat{I}_k^i(\mathbf{x})\right)}\right], \quad (4.34)$$

where $\widehat{I}_k^i$ is the i-exposed intensity fusion (in Eq. 4.9) and $\mathcal{N}$ is the kernel (Eq. 4.31) used for regression (in Fig. 4.20). The acquired spatial discrete and non-quantized radiance map, namely the HDR image $E : \mathbb{N}^2 \mapsto \mathbb{R}$ is a consistent (up-to-scale) manifold of the scene radiance L (see Sect. 4.7.5). In addition to the model calibration g_λ^M, the consistency of the resulting HDR images depends on: (i) The proper selection of integration times. (ii) The amount of images to be fused. (iii) The precise timing control. Therefore, a detailed analysis of the shutter control is presented (in Sect. 4.7.6). Afterwards, in Sect. 4.7.6 a novel method is presented for adaptively selecting the amount of exposures and their particular integration times for optimal radiance sampling. Finally, the robust and accurate image acquisition of various experimental scenarios is presented.

4.7.6 Exposure Control

Embodiment Aspects

When capturing the Wycoff set of a scene, the integration time Δt_i is controlled according to the camera specification [54] (see Fig. 4.4). However, there can be issues when using different camera models or even different firmware versions of the same camera (Figs. 4.22 and 4.23).

- **Exposure stability**: The first problem occurs when the integration time is not consistent for the whole image. This happens systematically when using rolling shutter sensors or (in case of global shutter) it may occur while dynamically changing the exposure settings. This issue is anticipated by the storage trigger. This trigger is fired by means of exposure stability analysis (see Fig. 4.24). This analysis takes also into account the latency[10] of the capture system. The exposure stability index is expressed as $S_i(t) := \frac{1}{wh} \sum_{\mathbf{x} \in \Omega} I_t^i(\mathbf{x})$. It integrates the image intensities while controlling the integration time Δt_i. Its differential analysis $\delta(S_i(t))/\delta t$ aids to determine reliable capture intervals D_j. Without this analysis, the storage trigger could be eventually fired within unstable intervals, for example, the instantaneous flickering between t_1 and t_4 with the peak at S_2 (see the zoom in Fig. 4.24b).
- **Indexed integration time**: Cameras supporting the specification [48] provide an indexed set of exposure times $\mathcal{E}$. However, depending on the particular camera and frame rate, the indexing integration times are not always regularly distributed (see Fig. 4.25). Without this consideration, the synthesis (Eq. 4.34) is corrupted.

Scene Aspects

Each scene has a singular radiance distribution depending on the lighting, materials and viewpoint. In order to optimally acquire the visual manifold of each scene, the humanoid robot can use the proposed method in [49] to fuse every available exposure (indexed integration time) of the camera to create a highly accurate HDR image. However, in practice this is not desirable nor feasible for online applications. The limitation is the long period of time (in the order of minutes) necessary to complete the HDR image acquisition. In this process, the amount and distribution of the exposures plays a critical role. Slightly differing integration times capture *"almost"* the same radiance segment. This redundancy can be removed without negative effects[11] during the HDR synthesis. Based on the radiometric calibration

[10]Irregular delays resulting from the non-real-time modular controller architecture.

[11]The noise reduction by integrating redundant exposures [44] is not necessary in the pipeline (see Fig. 4.4) due to the *"noiseless"* images attained by the optimal temporal fusion [55] during the exposure bracketing.

Fig. 4.22 The HDR image E acquired with the proposed method. This representation is $\ln(E)$ compressed to allow a suitable visibility of the sensitivity and consistency of the acquisition method. This HDR image was captured by the humanoid robot ARMAR-IIIa with its left peripheral camera in the scene setup of Fig. 4.16. This screenshot supports the quality and applicability of the method

Fig. 4.23 The HDR image E acquired with the proposed method. All images (inclusive the Sect. 4.7.5) correspond to a single HDR image captured by the humanoid robot ARMAR-IIIa with its left peripheral camera in the scene setup of Fig. 4.16. The variation of the perspective displays the discussed regions in Sect. 4.7

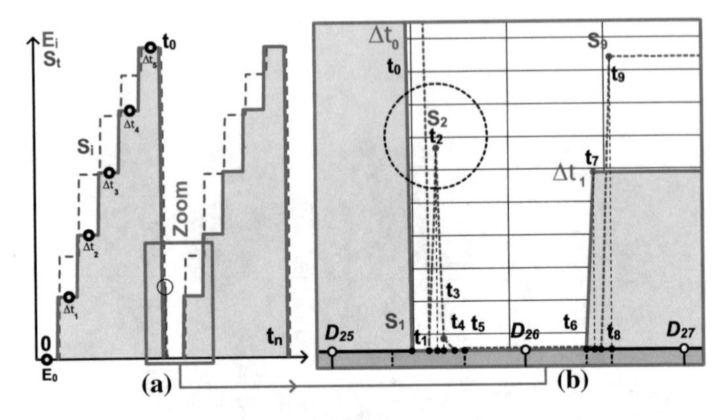

Fig. 4.24 **a** The exposure stability index S_t (blue dashed line) is closely correlated with the commanded integration time Δt_i (continuous red line). **b** The zoom-window shows the importance of the exposure stability analysis for the consistent capture of iso-exposed images I_k^i

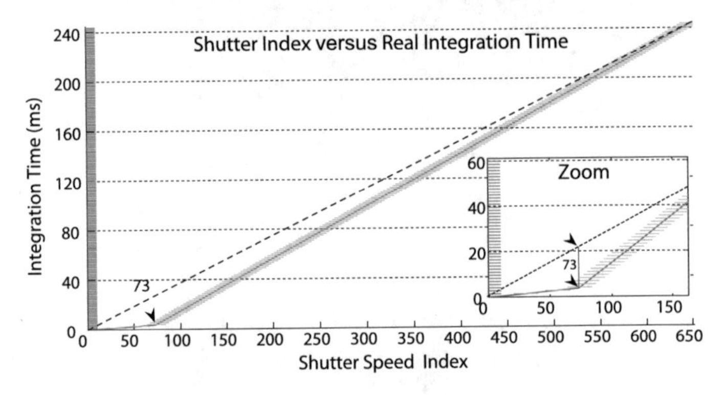

Fig. 4.25 The irregular distribution of the integration times (indexed integration times bend at index 73) according to the values accessible from the absolute registers of the camera interface (see [54])

model and the exposure granularity of a particular camera, it is possible to determine the minimal set of exposures $\mathcal{E}_{\min} \subsetneq \mathcal{E}$ necessary to sample the complete reachable radiance range of the sensor. When using a Bayer pattern camera, this minimal set of exposures depends on the radiometric calibration models ($g^M_{\lambda,R}$, $g^M_{\lambda,G}$ and $g^M_{\lambda,B}$) of the color channels, the regression kernel $\mathcal{N}$ used during calibration.

Sampling Density

Section 4.7.2 presents the reciprocity concept emphasizing the fact that the intensity extrema of the images do not hold the reciprocity principle. The comparative results of the proposed calibration model (in Fig. 4.20) experimentally support these observations. Despite the attained continuous and reciprocity consistent calibration model for HDR image synthesis, the physical nature of the sensor still produces unreliable measurements at the intensity extrema. Therefore, the kernel weighting $\mathcal{N}$ compensates these effects by gradually disregarding these intensity regions. By considering the calibration interval (Eq. 4.33), the weighting associated with the pixel location $\mathbf{x}$ resulting from the image capture with integration time Δt_i is expressed as

$$\kappa(\mathbf{x}, \Delta t_i) = \begin{cases} \mathcal{N}\left(\widehat{I}^i(\mathbf{x})\right), & \text{if } \mathbf{max}\left(\frac{-g_\lambda^L[1,2]}{g_\lambda^L[1,1]}, 1\right) < \widehat{I}^i(\mathbf{x}) < (2^m - 1) \\ 0, & \text{else.} \end{cases} \tag{4.35}$$

This piecewise weighting is also a density indicator of the representativeness of the value $\widehat{I}^i(\mathbf{x})$ at the exposure Δt_i. Ideally, all pixels in the image have to be (at least once) sampled with the unitary density, namely

$$\forall \mathbf{x} \in \Omega, \ \exists \Delta \hat{t}_i \in \mathcal{E} : \kappa(\mathbf{x}, \Delta \hat{t}_i) = 1. \tag{4.36}$$

However, because the radiance range can be rather wide and the pixel intensities are continuously and arbitrary spread along the sensor irradiance, the required amount of exposures is impractical large in terms scope and granularity. The required exposure set may have a cardinality larger than the cardinality of the available exposure set $\mathcal{E}$. Nevertheless, the accumulated sampling contributions attained by the exposure times ($\Delta t_j \approx \Delta \hat{t}_i$) are close enough to the ideal exposure time $\Delta \hat{t}_i$. These exposures provide the necessary information to approximate the ideal radiance sampling. This accumulated collection of sampling contributions (in terms of kernel weighting) is called radiance *sampling density* and is denoted as

$$\delta_\Phi(\mathbf{x}) := \sum_i^n \kappa(\mathbf{x}, \Delta t_i). \tag{4.37}$$

Based on this concept, the minimal set of exposures $\mathcal{E}_{\text{min}}$ ensuring the sampling density $\delta_\Phi(\mathbf{x}) > \kappa_m$, $\forall \mathbf{x} \in \Omega$ is the key to reduce the total visual manifold capturing time without decreasing the HDR image quality. Furthermore, the minimal set $\mathcal{E}_{\text{min}}$ is also subject to

$$\forall \mathbf{x} \in \Omega, \ \exists \Delta \check{t}_i \in \mathcal{E}_{\text{min}} \Rightarrow \kappa(\mathbf{x}, \Delta \check{t}_i) \geq \kappa_0. \tag{4.38}$$

This guarantees the minimal exposure overlapping (see Sect. 4.7.6). If all pixels captured from a scene cannot be sampled with the minimal sampling density κ_m, it can be affirmed that the lighting conditions are beyond the physical sensor capabilities. This is an important fact for reasoning and planning for robots.

Combined Radiance-Exposure Model

The radiometric image formation process (expressed by Eq. 4.25) and the radiometric calibration model (stated in Eq. 4.32) can be consistently combined

$$g(I^i) = \overbrace{\ln E + \ln \Delta t_i}^{\text{image exposure}} \cong \overbrace{\ln\left(I^i \cdot g^L_\lambda[1,1]) + g^L_\lambda[1,2]\right)}^{\text{radiometric calibration model}} = g^M_\lambda(I^i),$$
$$E \cong I^i \cdot \frac{g^L_\lambda[1,1]}{\Delta t_i} + \frac{g^L_\lambda[1,2]}{\Delta t_i}. \tag{4.39}$$

This linear[12] model associates the intensity values I^i according to the integration time Δt_i by means of the radiometric calibration model g^M_λ. Notice that Eq. 4.39 may yield to its direct and ideal applications for image synthesis. However, its straightforward application produces severe artifacts due to the broken reciprocity effects of the physical sensor. These effects have to be processed by kernel weighted fusion of various exposures (in the homomorphic domain as presented in Eq. 4.34).

This model depicts the relationship between radiance and exposure. It can be exploited to delineate the boundary curves of the radiance as functions of the integration time. This merges the radiometric calibration with the available integration times. Thus, it allows to determine the quantity and location of exposures necessary to extract the minimal exposure set $\mathcal{E}_{\min}$.

Full Reachable Radiance through Minimal Exposure Set

The extraction of the minimal exposure set $\mathcal{E}_{\min}$ requires a unified representation of the combined radiance-exposure model (Eq. 4.39), the valid interval of the calibration model (Eq. 4.33) and the sampling density (Eq. 4.37). The estimation of the set $\mathcal{E}_{\min}$ must also consider that Bayer pattern cameras have three different radiometric calibrations.

Bounding Curves

The radiometric calibration is defined only within the interval expressed in Eq. 4.33. Because the scene radiance $L \geq 0$ cannot be negative (see Fig. 4.4), the minimal valid intensity value (for all three color channels) is called the *"lower integrable intensity"* and is denoted as

$$\eta := \mathbf{max}\left(\left\lceil \frac{-g^L_{\lambda,R}[1,2]}{g^L_{\lambda,R}[1,1]} \right\rceil, \left\lceil \frac{-g^L_{\lambda,G}[1,2]}{g^L_{\lambda,G}[1,1]} \right\rceil, \left\lceil \frac{-g^L_{\lambda,B}[1,2]}{g^L_{\lambda,B}[1,1]} \right\rceil, 1\right), \tag{4.40}$$

where $\eta \geq 1$ for Eq. 4.32 to be defined. This implies a *lower bounding curve* E^l for the sensor irradiance as function of the integration time

$$\overbrace{E^l(\Delta t_i)}^{\text{lower bounding curve}} = \overbrace{\left(\eta \cdot g^L_\lambda[1,1] + g^L_\lambda[1,2]\right)}^{\text{fixed lower exposure}} \overbrace{(\Delta t_i)^{-1}}^{\text{shutter speed}}. \tag{4.41}$$

In the same manner, there is an *upper bounding curve* E^u defined by the maximal intensity minus the *over exposition clipping margin* $\nu > 1$

[12] Suitable for sensors with *"quasilinear"* behavior in the homomorphic domain.

$$\overbrace{E^{\mathrm{u}}(\Delta t_i)}^{\text{upper bounding curve}} = \overbrace{\left((2^m - \nu) \cdot g_\lambda^L[1,1] + g_\lambda^L[1,2]\right)}^{\text{fixed upper exposure}} \overbrace{(\Delta t_i)^{-1}}^{\text{shutter speed}} . \tag{4.42}$$

Following this pattern, the ideally sampled intensity $I_{\mathrm{mid}} = \frac{1}{2}(2^m - 1)$ defines the curve E^{mid}. At this intensity (according to Eq. 4.37) the maximal sampling density is found. The curve is expressed as

$$\overbrace{E^{\mathrm{mid}}(\Delta t_i)}^{\text{optimal sampling curve}} = \overbrace{\left(\frac{1}{2}(2^m - 1) \cdot g_\lambda^L[1,1] + g_\lambda^L[1,2]\right)}^{\text{fixed central exposure}} \overbrace{(\Delta t_i)^{-1}}^{\text{shutter speed}} . \tag{4.43}$$

Equation 4.43 is called *optimal sampling curve* (see Fig. 4.26). Since the sampling density is defined by a Gaussian kernel, it is possible to obtain the envelope curves at the lower and upper quantiles of the density (each at the first and second standard deviations $E^{(\mathrm{mid}+\epsilon\sigma)} \mid \epsilon \in \{-2, -1, 1, 2\}$, see Fig. 4.26). These curves contain the radiance regions for scene sampling with reference densities by the 68-95-99.7 rule [56].

Calibrated Exposures

The previous bounding and reference curves describe the continuous range which can be sampled depending on the integration time. Hence, each integration time available on the camera has an associated irradiance range. In Fig. 4.26 (at marker 1), the shortest available integration time $\Delta t_1 = 3.0994415 \mu s$ captures the highest reachable irradiance. The maximal intensity value $(2^m - 1) = 255$ @ $(m = 8)$ obtained within the time interval Δt_1 is produced by the sensor irradiance $g_\lambda^L = [0.00811, -0.0289]^T$

$$E((2^m - 1), \Delta t_1) = (2^m - 1) \cdot g_\lambda^L[1,1] + g_\lambda^L[1,2])(\Delta t_1)^{-1} = 657,908.9$$

Meanwhile its lower reachable intensity η corresponds to the irradiance $E(\eta, \Delta t_1) = 2,616.6$. Note the nonlinear behavior at the middle intensity value $E(\frac{1}{2}(2^m - 1), \Delta t_1) = 324,292.3$. Unfortunately, these extrema values are usually corrupted due to the broken reciprocity in physical sensors (see Sect. 4.7.2). In order to address this, the kernel weighting in terms of sampling density (Eq. 4.37) enables the integration of the information in a robust manner (Sect. 4.7.5). This means, beyond the cut-off sampling density κ the irradiance cannot be robustly captured. For example, the region within the cut-off sampling density $\kappa = 0.001$ is bounded with the intensities $[16, 239]$ (see Fig. 4.26 at (1)). By applying kernel weighting, the sensor irradiance range can be robustly sensed to $[32015.2, 616569.4]$ at Δt_1 from a larger (only ideal) range of $[2616.6, 657908.9]$. The kernel integrated range is only 89.2% of the ideal capacity of the sensor. The same situation is found at the longest integration time $\Delta t_{651} = 24.491072$ ms (see Fig. 4.26 at marker 2). In Fig. 4.26 (at marker 3), the lower irradiance captured by one exposure with the configuration ($\Delta t_{217} = 6.359$ @ $\kappa = 0.001$) is denoted as $E^{(\kappa=0.001, I=16)} = 15.6$, whereas the upper reachable irradiance is denoted as $E^{(\kappa=0.001, I=239)} = 300.5$. This range (shown in Fig. 4.26 at marker 4) depends on both the radiometric calibration and the selected sampling density

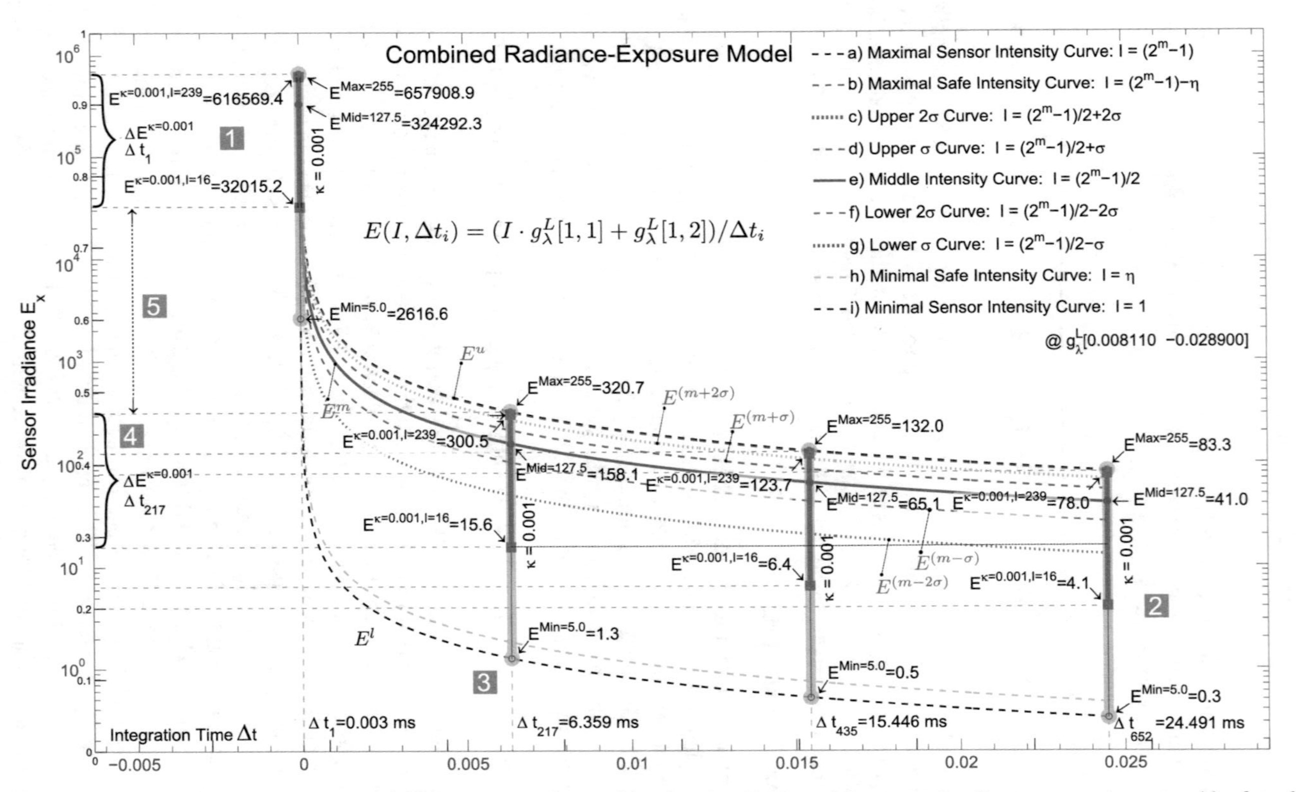

Fig. 4.26 The combined radiance-exposure model. This representation enables the visualization of the sensor irradiance segments captured by four different exposure examples

cut-off. Since the radiometric calibration is constant, the variation of the sampling density cut-off narrows or expands the irradiance range. The lower the cut-off the wider the range. Accordingly, with a lower cut-off less exposures are needed to cover the whole irrandiance range of the camera sensor. However, a low cut-off would also produce detrimental artifact due to the intrinsic broken reciprocity. Hence, the cut-off selection has to ensure the whole sensor irradiance domain with (at least) one exposure with the minimal sampling density κ_m.

The overall irradiance range captured with all available integration times (using a low cut-off sampling density $\kappa = 0.001$) is $[4.1, 616569.4]$. Hence, the intensity value (per channel) requires at least

$$\lceil \log_2(616,569.4 - 4.1) \rceil = 20 \text{ bits}$$

for its integer representation. For this reason, the radiance representation in this work are real values. Therefore, a floating point representation with 32 bit per channel and 12 bytes per color pixel (after demosaicing in Sect. 5.2) is used. This large computational representation ensures the forward compatibility for high speed cameras with shorter integration times and higher resolution analog-to-digital converters.

Minimal Calibrated Exposure Set

Algorithm 7 describes the applied method to select the minimal exposure set $\mathcal{E}_{\min}$. The strategy is to select the exposures producing the minimal sensor irradiance overlapping. This process starts from the shortest integration time towards the longest integration time. This is addressed in this manner because the short integration exposures are less redundant than the long integration exposures (see the resulting set $\mathcal{E}_{\min}$ in Fig. 4.27).

Algorithm 7 Extract-Minimal-Exposure-Set, $EMES(\kappa,\sigma,g^L_\lambda,\eta,\nu,\mathcal{E})$

Require: $\kappa > 0$ **and** $|\mathcal{E}| > 1$
1: $\mathcal{E}_{\min} \leftarrow \Delta t_1$
2: $I_l \leftarrow \mathbf{max}(\frac{1}{2}(2^m - 1) - log(\kappa\sigma^2)^{\frac{1}{2}}, \eta)$, $I_u \leftarrow \mathbf{min}(\frac{1}{2}(2^m - 1) + log(\kappa\sigma^2)^{\frac{1}{2}}, \nu)$
3: **for** $i = 2$ **to** $|\mathcal{E}|$ **do**
4: $\quad E^l_i \leftarrow (I_l g^L_\lambda[1,1] + g^L_\lambda[1,2])/\Delta t_i$, $E^u_i \leftarrow (I_u g^L_\lambda[1,1] + g^L_\lambda[1,2])/\Delta t_i$
5: $\quad O_{\min} \leftarrow \infty$, $k = 0$
6: $\quad$ **for** $j = (i+1)$ **to** $|\mathcal{E}|$ **do**
7: $\quad\quad E^l_j \leftarrow (I_l g^L_\lambda[1,1] + g^L_\lambda[1,2])/\Delta t_j$, $E^u_j \leftarrow (I_u g^L_\lambda[1,1] + g^L_\lambda[1,2])/\Delta t_j$
8: $\quad\quad$ **if** $(E^u_j > E^l_i) \wedge ((E^u_j - E^l_i) < O_{\min})$ **then**
9: $\quad\quad\quad O_{\min} \leftarrow (E^u_j - E^l_i)$, $k \leftarrow j$
10: $\quad\quad$ **end if**
11: $\quad$ **end for**
12: $\quad$ **if** $(k > 0)$ **then**
13: $\quad\quad \mathcal{E}_{\min} \leftarrow \mathcal{E}_{\min} \cup \Delta t_k$
14: $\quad$ **end if**
15: **end for**
16: **return** $\mathcal{E}_{\min}$

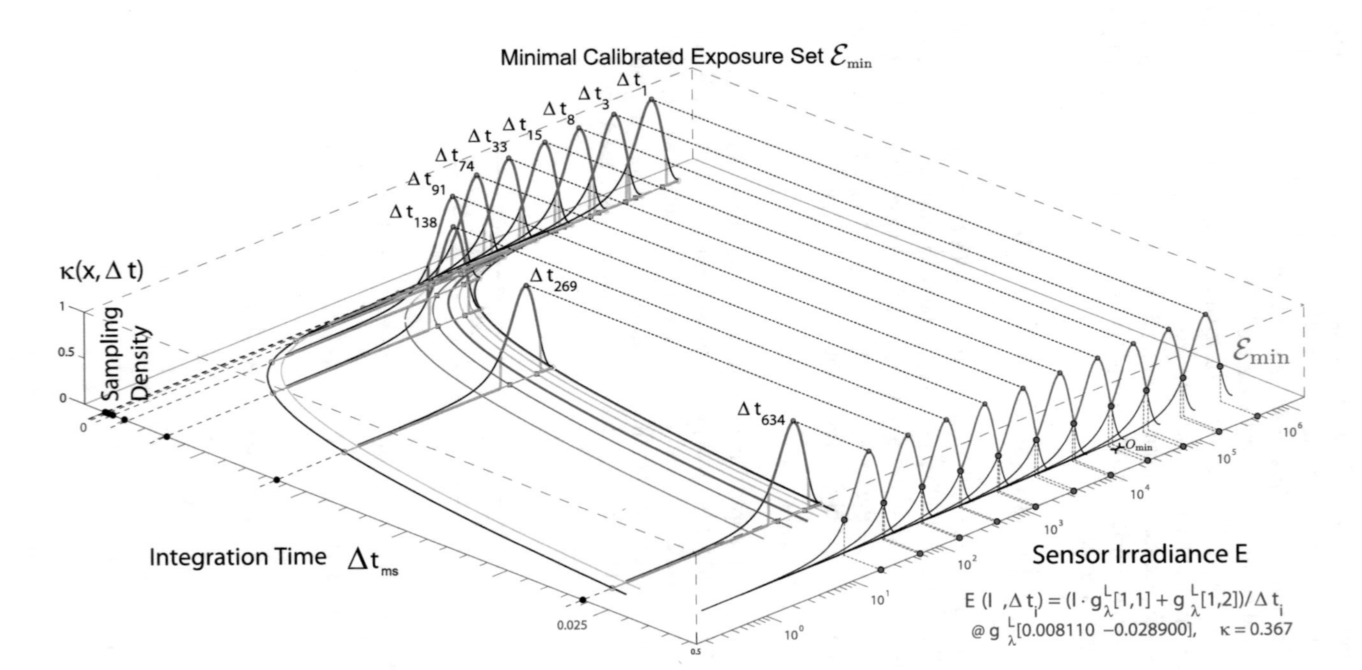

Fig. 4.27 The minimal calibrated exposure set containing 11 exposures out of 651. The set is $\mathcal{E}_{\min} := \{\Delta t_1 = 0.0031, \Delta t_3 = 0.0083, \Delta t_8 = 0.0212, \Delta t_{15} = 0.0550, \Delta t_{33} = 0.1488, \Delta t_{74} = 0.3989, \Delta t_{91} = 1.1075, \Delta t_{138} = 3.0665, \Delta t_{269} = 8.5269, \Delta t_{634} = 23.7408, \Delta t^*_{651} = 24.4911\}$ in ms. The exposure Δt_{651} is not depicted to avoid visual overlapping. This results from Algorithm 7 with $\kappa_m = 0.3671381$. This particular cut-off is determined by the larger gap ($O_{\min}$ in lines 7–12 of Algorithm 7) between two consecutive exposures using the kernel weighting of Eq. 4.37. A lower cut-off $\kappa < \kappa_m$ would generate either sampling holes in the irradiance domain or the minimal sampling density could not be thoroughly ensured. Therefore, the minimal exposure set is fully determined by both the radiometric calibration and the integration time granularity of the camera

Fig. 4.28 Everyday scene where the visual perception of the humanoid robot must recognize environmental elements for attaining its own 6D-pose. This viewpoint contains considerably wide dynamic range produced by the large radiance difference between the ceiling lamps and the oven

4.8 Experimental Evaluation: Dynamic Range

The coordinated integration of both methods (Sects. 4.5 and 4.7) enables humanoid robots to acquire the visual manifold in a robust and high quality manner (see Figs. 4.28, 4.29 and 4.30). This high signal-to-noise ratio results from integrating multiple images while precisely controlling the exposure (see Sect. 4.7.6). The minimal amount of exposures is soundly attained by two considerations: (i) The minimal exposure set is extracted based on the improved radiometric calibration and the exposure granularity of the cameras. (ii) At each exposure configuration, a temporal fusion is conducted according to the convergence trade-off χ_k (see Fig. 4.29). This determines the minimal amount of images to be captured while simultaneously avoiding redundant sampling and ensuring high quality.

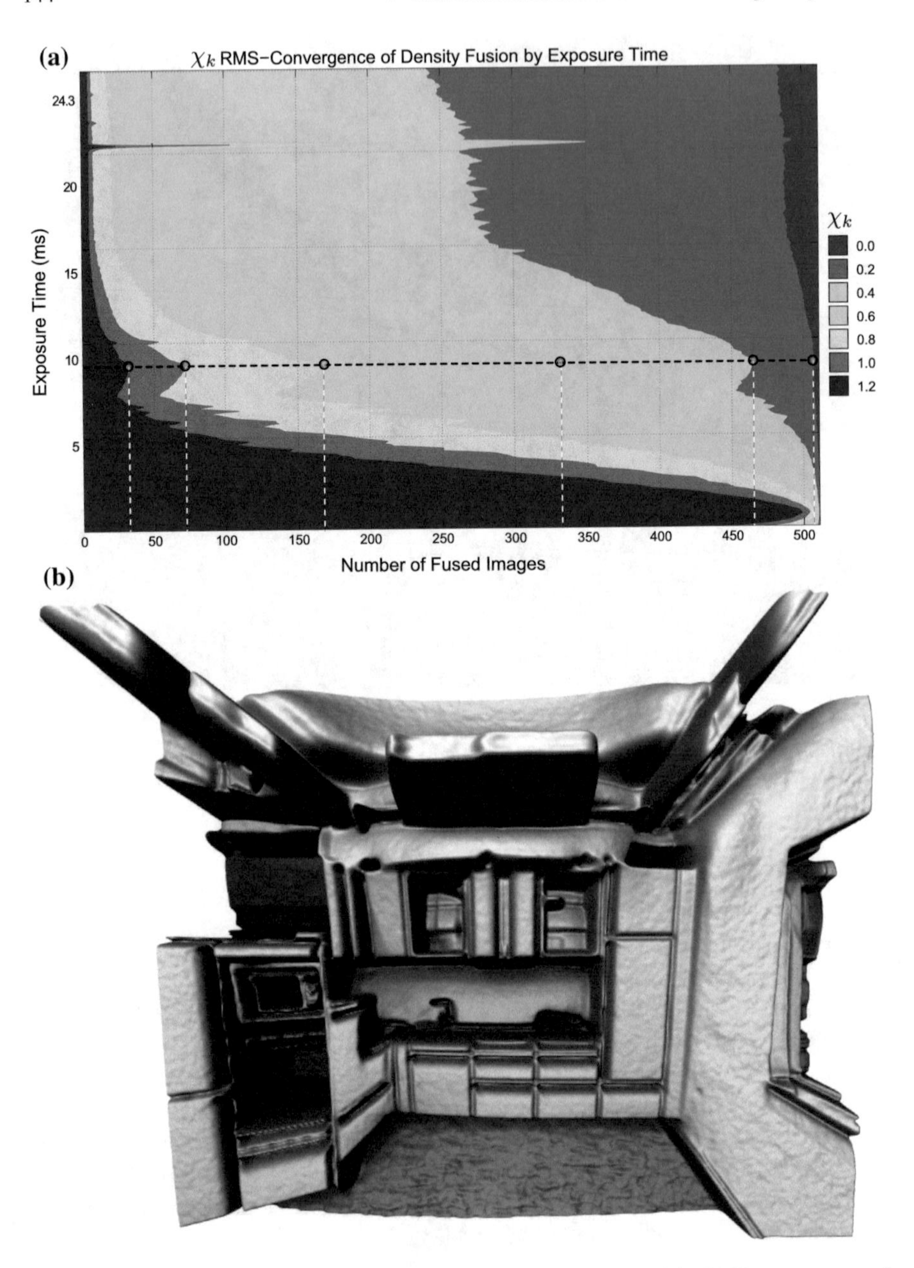

Fig. 4.29 Visual manifold acquisition. **a** The plot shows the analysis of the RMS-convergence of temporal fusion χ_k (Eq. 4.18) depending on integration times. The dotted line at Δt_{269} shows five different levels of quality with their corresponding amount of images. **b** The lower HDR mesh is the $\ln(E)$ compressed representation of the radiance scene acquired with VGA resolution using the 4 mm lens (in Fig. 4.28, see also the integrated HDR acquisition pipeline in Fig. 4.4)

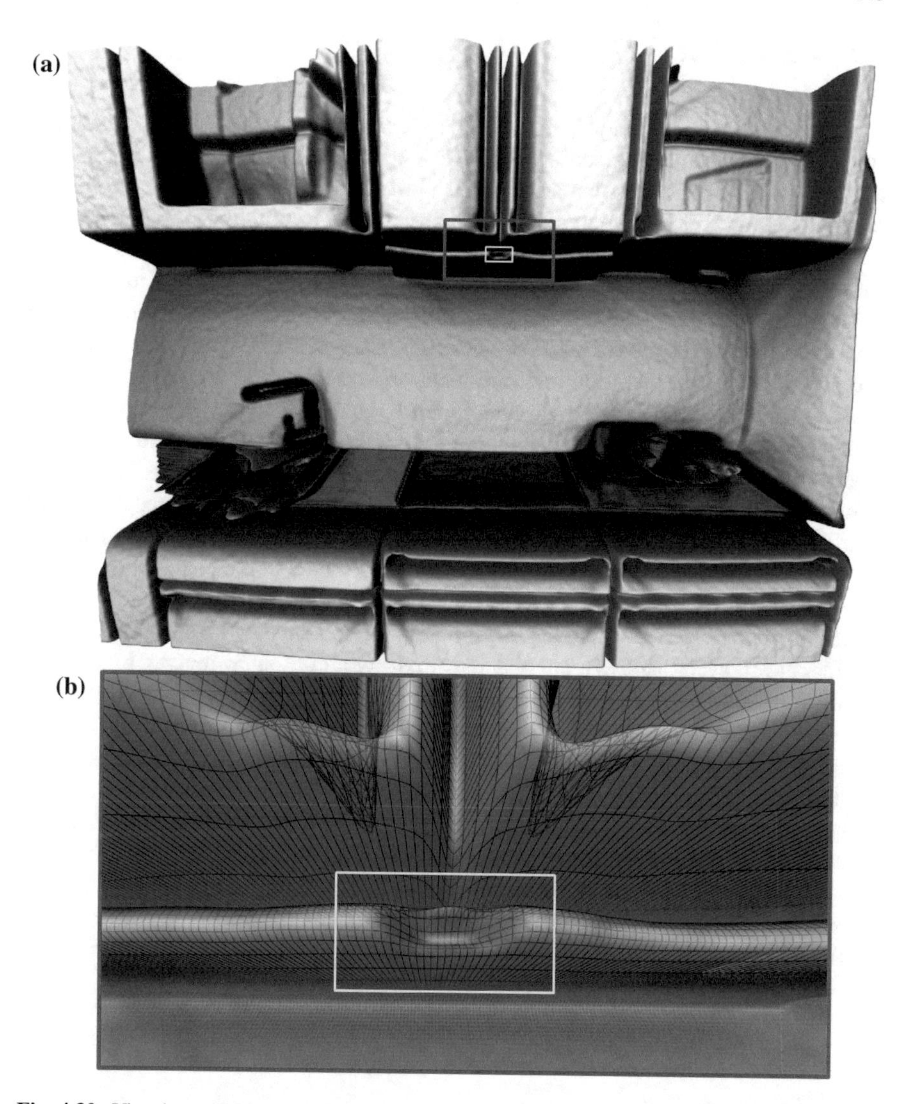

Fig. 4.30 Visual manifold acquisition detail. **a** The HDR mesh $\ln(E)$ of the region marked with a rectangle in Fig. 4.28. This visual manifold was acquired with a 12 mm lens. Using a large focal length significantly improves the spatial resolution. **b** The red rectangle is shown in the zoom illustrating the high quality of the synergistic integration of both methods

4.9 Discussion

In the first part of this chapter, a new method based on multiple image fusion for robust visual manifold acquisition was presented and evaluated. The method is suitable for complex robot systems including humanoid robots. The acquired images resulting from this fusion method conveniently overcome both the embodiment limitations of the humanoid robots and the unsuitable environmental conditions. Because the optimal estimation of the irradiance value is done per pixel, the method has been tested with Bayer pattern images, improving the resulting color images after a semi-continuous demosaicing (see Sect. 5.2). Categorically, the fusion method is a temporal image registration and optimization which clearly improves all visual sensing tasks by providing stable and superior quality images. The experimental evaluation clearly supports the benefits of the method by providing up to a 40.05% segmentation improvement in terms of stability compared to the results attained by using directly captured images. Moreover, the edge extraction improves by up to a 23.81% compared to directly captured images. This method enhances the image for more representativeness and repeatability of the extracted features allowing a wider applicability of all robot vision methods in semi-dynamic scenarios.

In the second part of this chapter, the *"noiseless"* images captured by temporal fusion are systematically merged into high-dynamic-range images by optimal exposure bracketing. This dynamic range expansion is precisely archived through the improved radiometric calibration and meticulously exposure control. The proposed radiometric calibration model properly holds the reciprocity principle even at the intensity extrema with an efficient computational representation for continuous images. The synergistic coordination of noise removal and dynamic range enhancement arises at the convergence behavior analysis. Based on these elements, the optimal combination of temporal fusion and exposure bracketing is properly controlled. This is concretely reflected in the high quality images for feature extraction with minimal acquisition time. These observations are supported by experimental evaluation.

References

1. Inaba, M., T. Igarashi, S. Kagami, and H. Inoue. 1996. A 35 DOF Humanoid that Can Coordinate Arms and Legs in Standing up, Reaching and Grasping an Object. In *IEEE-RSJ International Conference on Intelligent Robots and Systems*, vol. 1, 29–36.
2. Scassellati, B. 1998. A Binocular, Foveated Active Vision System. Technical report, MIT Artificial Intelligence Lab, Cambridge, MA, USA.
3. Welke, K., T. Asfour, and R. Dillmann. 2009. Active Multi-view Object Search on a Humanoid Head. In *IEEE International Conference on Robotics and Automation*, 417–423.
4. Asfour, T., K. Regenstein, P. Azad, J. Schröder, A. Bierbaum, N. Vahrenkamp, and R. Dillmann. 2006. ARMAR-III: An Integrated Humanoid Platform for Sensory-Motor Control. In *IEEE-RAS International Conference on Humanoid Robots*, 169–175.
5. Asfour, T., K. Welke, P. Azad, A. Ude, and R. Dillmann. 2008. The Karlsruhe Humanoid Head. In *IEEE-RAS International Conference on Humanoid Robots*, 447–453.

6. Bayer, B. 1976. Color Imaging Array. US Patent US3 971 065 A.
7. Rao, D., and P. Panduranga. 2006. A Survey on Image Enhancement Techniques: Classical Spatial Filter, Neural Network, Cellular Neural Network, and Fuzzy Filter. In *IEEE International Conference on Industrial Technology*, 2821–2826.
8. Ginneken, B., and A. Mendrik. 2006. Image Denoising with K-nearest Neighbor and Support Vector Regression. In *International Conference on Pattern Recognition*, vol. 3, 603–606.
9. Buades, A., B. Coll, and J.M. Morel. 2004. On Image Denoising Methods. Technical Note, Centre de Mathematiques et de Leurs Applications.
10. Mahmoudi, M., and G. Sapiro. 2005. Fast Image and Video Denoising via Nonlocal Means of Similar Neighborhoods. *IEEE Signal Processing Letters* 12 (12): 839–842.
11. Tasdizen, T. 2008. Principal Components for Non-local Means Image Denoising. In *IEEE International Conference on Image Processing* , 1728 –1731.
12. Kharlamov, A., and V. Podlozhnyuk. 2007. Image Denoising, NVIDIA Inc., Technical Report.
13. Dabov, K., R. Foi, V. Katkovnik, and K. Egiazarian. 2009. BM3D Image Denoising with Shape-Adaptive Principal Component Analysis. In *Workshop on Signal Processing with Adaptive Sparse Structured Representations*.
14. Irshad, H., M. Kamran, A. Siddiqui, and A. Hussain. 2009. Image Fusion Using Computational Intelligence: A Survey. In *International Conference on Environmental and Computer Science*, 128–132.
15. Battiato, S., A. Bruna, and G. Puglisi. 2010. A Robust Block-Based Image/Video Registration Approach for Mobile Imaging Devices. *IEEE Transactions on Multimedia* 12 (7): 622–635.
16. Glasner, D., S. Bagon, and M. Irani. 2009. Super-resolution from a Single Image. In *IEEE International Conference on Computer Vision*, 349–356.
17. Park, S.C., M.K. Park, and M.G. Kang. 2003. Super-resolution Image Reconstruction: A Technical Overview. *IEEE Signal Processing Magazine* 20 (3): 21–36.
18. Chen, C., L. Phen-Lan, and H. Po-Whei. 2010. A New Fusion Scheme for Multi-focus Images based on Dynamic Salient Weights on Discriminative Edge Points. In *International Conference on Machine Learning and Cybernetics*, 351–356.
19. Szeliski, R., M. Uyttendaele, and D. Steedly. 2011. Fast Poisson Blending using Multi-splines. In *IEEE International Conference on Computational Photography*, 1–8.
20. Pessoa, S., G. Moura, J. Lima, V. Teichrieb, and J. Kelner. 2010. Photorealistic Rendering for Augmented Reality: A Global Illumination and BRDF Solution. In *IEEE Virtual Reality Conference*, 3–10.
21. Chen, L., X. Wang, and X. Liang. 2010. An Effective Video Stitching Method. In *International Conference on Computer Design and Applications*, vol. 1, 297–301.
22. Gonzalez-Aguirre, D., and E. Bayro-Corrochano. 2006. A Geometric Approach for an Intuitive Perception System of Humanoids. *Intelligent Autonomous Systems* 9: 399–407.
23. Mann, S., and R.W. Picard. 1995. On Being 'undigital' With Digital Cameras: Extending Dynamic Range By Combining Differently Exposed Pictures. In *Annual Conference; Society for Imaging Science and Technology*, 442–448.
24. Marks, R.J. 2009. *Handbook of Fourier Analysis and Its Applications*. USA: Oxford University Press. ISBN 978-0195335927.
25. Rice J. 2006. *Mathematical Statistics and Data Analysis*. Duxbury Press. ISBN 978-0534399429.
26. Duda, R., P. Hart, and D. Stork. 2001. *Pattern Classification*, 2nd ed. New York: Wiley. ISBN 978-0471056690.
27. Elgammal, A., R. Duraiswami, and L. Davis. 2003. Efficient Kernel Density Estimation using the Fast Gauss Transform with Applications to Color Modeling and Tracking. In *IEEE Transactions on Pattern Analysis and Machine Intelligence* 25 (11): 1499–1504.
28. Härdle, W., M. Müller, S. Sperlich, and A. Werwatz. 2004. *Nonparametric and Semiparametric Models*. New York: Springer.
29. Qi, Z., and R. Hong-e. 2009. A New Idea for Color Annual Ring Image Segmentation Adaptive Region Growing Algorithm. In *International Conference on Information Engineering and Computer Science*, 1–3.

30. Grigorescu, C., N. Petkov, and M. Westenberg. 2003. Contour Detection based on Nonclassical Receptive Field Inhibition. *IEEE Transactions on Image Processing* 12 (7): 729–739.
31. Healey, G., and R. Kondepudy. 1994. Radiometric CCD Camera Calibration and Noise Estimation. *IEEE Transactions on Pattern Analysis and Machine Intelligence* 16 (3): 267–276.
32. Dominguez-Tejera, J. 2012. GPU-Based Low-Level Image Processing for Object Recognition using HDR Images. Master's thesis, KIT, Karlsruhe Institute of Technology, Computer Science Faculty, Institute for Anthropomatics.
33. Okada, K., M. Kojima, S. Tokutsu, Y. Mori, T. Maki, and M. Inaba. 2008. Task Guided Attention Control and Visual Verification in Tea Serving by the Daily Assistive Humanoid HRP2JSK. In *IEEE-RSJ International Conference on Intelligent Robots and Systems*, 1551–1557.
34. Wieland, S., D. Gonzalez-Aguirre, T. Asfour, and R. Dillmann. 2009. Combining Force and Visual Feedback for Physical Interaction Tasks in Humanoid Robots. In *IEEE-RAS International Conference on Humanoid Robots*, 439–446.
35. Gonzalez-Aguirre, D., T. Asfour, E. Bayro-Corrochano, and R. Dillmann. 2008. Model-based Visual Self-localization Using Geometry and Graphs. In *International Conference on Pattern Recognition*, 1–5.
36. Asfour, T., P. Azad, N. Vahrenkamp, K. Regenstein, A. Bierbaum, K. Welke, J. Schröder, and R. Dillmann. 2008. Toward Humanoid Manipulation in Human-centred Environments. *Robotics and Autonomous Systems* 56 (1): 54–65.
37. Tokutsu, S., K. Okada, and M. Inaba. 2009. Environment Situation Reasoning Integrating Human Recognition and Life Sound Recognition using DBN. In *IEEE International Symposium on Robot and Human Interactive Communication*, 744–750.
38. Gonzalez-Aguirre, D., S. Wieland, T. Asfour, and R. Dillmann. 2009. On Environmental Model-Based Visual Perception for Humanoids. In *Progress in Pattern Recognition, Image Analysis, Computer Vision, and Applications*, ed. E. Bayro-Corrochano, and J.-O. Eklundh, 901–909. Lecture Notes in Computer Science. Berlin: Springer.
39. Lindeberg, T. 1996. Edge Detection and Ridge Detection with Automatic Scale Selection. *International Journal of Computer Vision* 30 (2): 465–470.
40. Welke, K., T. Asfour, and R. Dillmann. 2009. Bayesian Visual Feature Integration with Saccadic Eye Movements. In *IEEE-RAS International Conference on Humanoid Robots*, 256–262.
41. Vernon, D., G. Metta, and G. Sandini. 2007. The iCub Cognitive Architecture: Interactive Development in a Humanoid Robot. In *IEEE International Conference on Development and Learning*, 122–127.
42. Metta, G., L. Natale, F. Nori, and G. Sandini. 2011. The iCub Project: An Open Source Platform for Research in Embodied Cognition. In *IEEE Workshop on Advanced Robotics and its Social Impacts*, 24–26.
43. Grossberg, M.D., and S.K. Nayar. 2002. What Can Be Known about the Radiometric Response Function from Images. In *IEEE Conference on Computer Vision and Pattern Recognition*, vol. 2, 602–609.
44. Debevec, P., and J. Malik. 1997. Recovering High Dynamic Range Radiance Maps from Photographs. In *Annual conference on Computer Graphics and Interactive Techniques*, 369–378.
45. Mitsunaga, T., and S. Nayar. 1999. Radiometric Self Calibration. In *IEEE Computer Society Conference on Computer Vision and Pattern Recognition*, vol. 2, 637–663.
46. Krawczyk, G., M. Goesele, and H.-P. Seidel. 2005. Photometric Calibration of High Dynamic Range Cameras, Max-Planck-Institut für Informatik, 66123 Saarbrücken, Germany, Research Report.
47. Lin, S., J. Gu, S. Yamazaki, and H.-Y. Shum. 2004. Radiometric Calibration from a Single Image. In *IEEE Computer Society Conference on Computer Vision and Pattern Recognition*, vol. 2, 938–945.
48. IIDC2. 2012. *Digital Camera Control Specification Ver.1.0.0*, Japan Industrial Imaging Association Standard JIIA CP-001-2011/1394 Trade Association Specification TS2011001, Japan.
49. Gonzalez-Aguirre, D., T. Asfour, and R. Dillmann. 2010. Eccentricity Edge-Graphs from HDR Images for Object Recognition by Humanoid Robots. In *IEEE-RAS International Conference on Humanoid Robots*, 144 –151.

50. Theuwissen, A. 1995. *Solid-state Imaging with Charge Coupled Devices* Dordrecht: Kluwer Academic Publishers. ISBN 0-7923-3456-6.
51. KIT. 2011. Karlsruhe Institute of Technology, Computer Science Faculty, Institute for Anthropomatics, The Integrating Vision Toolkit. http://ivt.sourceforge.net.
52. Mackiewich, B. 1995. Intracranial Boundary Detection and Radio Frequency Correction in Magnetic Resonance Images. Master's thesis, Simon Fraser University.
53. Madisetti, V., and D. Williams. 2009. *The Digital Signal Processing Handbook*. CRC Press. ISBN 9781420045635.
54. PTGrey. 2008. *Dragonfly Technical Reference Manual*. Accessed 5 Aug 2008.
55. Gonzalez-Aguirre, D., T. Asfour, and R. Dillmann. 2011. Robust Image Acquisition for Vision-Model Coupling by Humanoid Robots. In *IAPR-Conference on Machine Vision Applications*, 557–561.
56. Zwillinger, D., and S. Kokoska. 2000. *CRC Standard Probability and Statistics Tables and Formulae*. Boca Raton: CRC Press.

Chapter 5
Environmental Visual Object Recognition

The visual manifold acquisition (presented in Chap. 4) supplies high quality signals necessary to consistently extract visual-features for establishing the vision model coupling. The combined sensitivity and stability of the acquired HDR images provide a high degree of independence from photometric and radiometric effects compared to the straightforward process of capturing noisy LDR images. This occurs because HDR images properly capture the ubiquitous structural visual-features typically observed in human centered environments.

In this work, the visual-feature representation is defined as: *"The sensor-dependent language which encompasses a formal description of the manifestations of the object shapes found in images acquired within the represented domain"* (see Chap. 3). Since the application domain is the visual perception for manipulation and grasping, the visual-feature representation has to be tightly related to the world model representation in terms of spatial metrics. This implies, the counterpart representation criteria used in the world model representation (see Sect. 3.1) have to be coherently considered and reflected in the visual-features. Therefore, the manifestation of the geometrical edges and vertices of the elements in the world are the first and most important source of information (see [1]) for geometric perception of the humanoid robots. However, a comprehensive structural visual-feature extraction such as edge and corner from images are not always plausible when using methods based solely on a single image cue (see discussion in [2]). These problems arise in various image contents. For example, when thin elements, small structures or dense clutter are present in the scene. Thus, additional image processing cues have to be incorporated for robust structural visual-feature extraction.

D. I. González Aguirre, *Visual Perception for Humanoid Robots*,
Cognitive Systems Monographs 38, https://doi.org/10.1007/978-3-319-97841-3_5

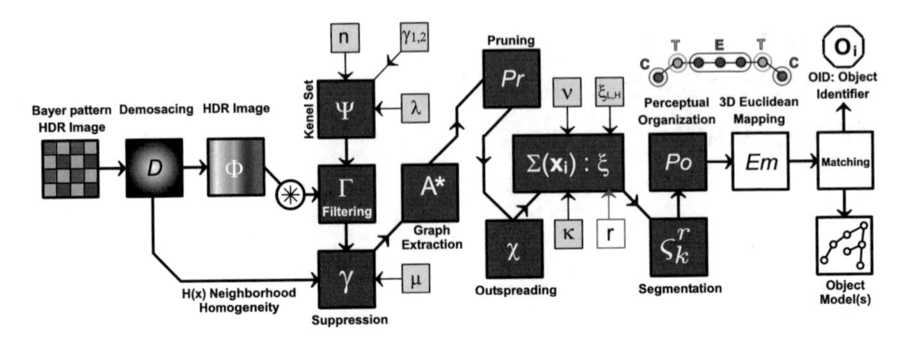

Fig. 5.1 Schematic representation of the multi-cue pipeline for visual object recognition. The three cues homogeneity, edge and rim are integrated in order to extract perceptually organized 3D geometric-primitives

5.1 Multi-cue Pipeline

In this chapter, a method is presented for extracting and integrating three image cues to obtain high representative visual-features (see schema in Fig. 5.1). The first cue is the *neighborhood homogeneity* (see Sect. 5.2) which is obtained as residual information of the adaptive demosaicing [3]. Later (in Sect. 5.4), the lens undistortion produce consistent images for euclidean metric extraction. Notice that before extracting the second and third cue, the spatial resolution of both images is virtually increased by means of bicubic interpolation [4]. This interpolation produces positive, non-evident and interesting effects for the next processing stages (see Sect. 5.5). Subsequently, a receptive radiance saliency function is computed based on [5] for the extraction of the second and third cues, namely the *continuous edges* and the *continuous phase rim* (see Sect. 5.6). The novel representation of both cues in a graph structure enables parameterless segmentation of geometric edges from the objects in the world. Later, the fusion of the information provided by the cues is discussed along the chapter. The resulting geometric-primitives properly reflect the physical structure of objects in the field of view and can cope with a variety of scenarios. The experimental evaluation of the method is presented in Sect. 5.8. Finally, the discussion of the pipeline is given in Sect. 5.9.

5.2 HDR Bayer Pattern Image Reconstruction

The acquired HDR images (in Sect. 4.8) are synthesized as Bayer pattern for subsequent reconstruction by demosaicing.[1] There are three reasons to proceed in this manner:

[1]Demosaicing is the reconstruction of a color image based on partial color samples from a sensor using a color filter array such as the wide spread Bayer pattern.

- **Radiometric aspects**: The radiometric calibration model is different for each color channel and eventually for each pixel (see Fig. 4.20).
- **Improved chromatic estimation**: The reconstruction of the color HDR image takes advantage of the wide dynamic range and high signal-to-noise ratio of the acquisition pipeline producing superior color approximations.
- **Computational efficiency**: The computational load necessary to generate the Bayer pattern HDR images is only one third of the load compared to the full RGB color HDR images. This implies no information loss in the pipeline because the color image reconstruction is done consistently (see Fig. 4.4).

The demosaicing problem is a well studied topic. Hence, there is a wide variety of methods (see [3, 6–8]) which can be properly compared according to their peak signal-to-noise ratio (see evaluations in [3, 8]). Figure 5.2 shows the quality of the method and its evaluation in [3]. Because this method is based on bilateral filtering (see [9]), the pixel-wise adaptive filtering kernel can be used to obtain a scalar indicator value which depicts the neighborhood homogeneity as follows.

First, there are different Bayer patterns $\mathcal{B} := \{GRBG, RGGB, BGGR, GBRG\}$ depending on the particular camera sensor. Hence, for a pixel at the location $\mathbf{x} \in \mathbb{N}^2$ in a Bayer pattern HDR image $E_{\mathcal{B}_i}(\mathbf{x}) \in \mathbb{R}$ is possible to determine its color component as a table called color-channel map $C(\mathcal{B}_i, \mathbf{x}) \in \{R, G, B\}$. Based on this table, the green channel image is estimated by means of the method [6]. This produces the first approximation image for the green channel image E_G.

Next, the adaptive demosacing algorithm exploits the first approximation of the green channel image E_G to determine the similarity of neighborhood pixels for a bilateral adaptive reconstruction E_G^a as

$$E_G^a(\mathbf{x}) = \frac{\sum_{\mathbf{x}_i \in \omega_s} \left\{ E_G(\mathbf{x}_i) \cdot \mathbf{exp}\left[\overbrace{\frac{(E_G(\mathbf{x}_i) - E_G(\mathbf{x}))^2}{-2\sigma_r^2}}^{\text{irradiance distance}} + \overbrace{\frac{||\mathbf{x}_i - \mathbf{x}||^2}{-2\sigma_s^2}}^{\text{spatial distance}} \right] \right\}}{\underbrace{\sum_{\mathbf{x}_i \in \omega_s} \underbrace{\mathbf{exp}\left(\frac{(E_G(\mathbf{x}_i) - E_G(\mathbf{x}))^2}{-2\sigma_r^2} + \frac{||\mathbf{x}_i - \mathbf{x}||^2}{-2\sigma_s^2} \right)}_{A_G(\mathbf{x},\mathbf{x}_i) \text{ adaptive kernel}}}_{H(\mathbf{x}) \text{ neighborhood homogeneity}}}, \tag{5.1}$$

where σ_r and σ_s represent the range and spatial bandwidth (standard deviation) of the Gaussian weighting kernel. Furthermore, the spatial standard deviation σ_s determines the cardinality of the pixel neighborhood $|\omega_s| = (\lceil 3\sigma_s \rceil + 1)^2$. By exploiting the adaptive kernel $A_G(\mathbf{x}, \mathbf{x}_i)$ of the green channel, the adaptive green E_G^a, red E_R^a and blue E_B^a color planes are adaptively estimated. The key difference to Eq. 5.1 is that for those pixel location where the channel is not available the bilateral kernel weighting is consider to be zero. The underlying assumption of this method is sup-

Fig. 5.2 Demosaicing methods for Bayer pattern images. The textured region is selected to clearly show the qualitative results. **a** The fast (1.2–2 ms) but naive bilinear interpolation exposes various detrimental effects, for example zipper effects, lost of sharpness and jagged edges [10]. **b** The demosaicing with edge direction detection [6] provides better results with relative fast performance (15–20 ms). **c** The adaptive demosaicing method [3] provides the best results of all previous methods. However, its time performance is the lowest (85–100 ms) due to the internal bilateral filtering

ported by the fact that the luminance is better (due to its location along the spectrum) estimated by the green channel. The resulting bilateral filtered values have a convenient edge preserving smoothing effect consequence of the adaptive kernel. This means, the resulting images are simultaneously reconstructed and smoothed without losing acutance.

Fig. 5.3 The globally normalized neighborhood homogeneity H' is a cue useful to determine whether a particular pixel $H'(\mathbf{x})$ is located within a homogeneous region or not. Notice that the homogeneity image has been displayed using a color map for a better visualization

5.3 Homogeneity Cue

In Eq. 5.1, the neighborhood homogeneity $H(\mathbf{x})$ is the product[2] of two kernels, the range kernel implemented by Gaussian metric of the irradiance distance and the spatial metric also implemented by a Gaussian kernel. The spatial Gaussian kernel solely depends on the pixel euclidean distance between the central pixel and the elements in the neighborhood. Hence, pixels located in regions with high variance produce low neighborhood homogeneity (see Fig. 5.3).

5.4 Full Image Lens Undistortion

The previous demosacing is the last step of the image acquisition. However, the resulting images exhibit lens distortion. This distortion has a dual nature, the so-called radial and tangential distortions. Both distortions depend on the lens transfer function and are usually highly correlated with the focal distance of the lens. The shorter the focal distance the stronger distortion of the images (see Fig. 5.4). The correction of the lens distortion is fundamental for the application of linear mapping models for the extraction of euclidean metric from images [11]. For example, when using epipolar geometry to establish correspondences and triangulation for 3D estimation [12]. The undistortion process is done for each camera and is based on both the pinhole camera model [11] and the lens distortion polynomial models [13] as follows. First,

[2]Factorized as a sum in the exponent of $A_G(\mathbf{x}, \mathbf{x}_i)$ in Eq. 5.1.

without lost of generality consider the camera at the origin of the coordinate system. A point $P = (x, y, z) \in \mathbb{R}^3$ in front of the camera ($z >$ M.O.D.[3]) is mapped to the image coordinates $p \in \mathbb{R}^2$ by the pinhole camera model with lens distortion, namely $\xi : \mathbb{R}^2 \mapsto \mathbb{R}^2$ as

$$\begin{aligned}
x' &= x/z, \; y' = y/z, \\
r &= \sqrt{x'^2 + y'^2}, \\
\delta &= 1 + k_1 r^2 + k_2 r^4 + k_3 r^6, \\
\tau_x &= 2 p_1 x' y' + p_2 (r^2 + 2x'^2), \\
\tau_y &= 2 p_2 x' y' + p_1 (r^2 + 2y'^2), \\
p &= [f_x (x'\delta + \tau_x) + c_x, \; f_y (y'\delta + \tau_y) + c_y]^T
\end{aligned} \tag{5.2}$$

where f_x and f_y are the focal distances in each direction for considering the case of non-square pixels of the CCD sensor [11]. In this context, δ is the radial distortion depending on the coefficients $k_1, k_2, k_3 \in \mathbb{R}$. The tangential distortion per axis τ_x and τ_y depend on the coefficients p_1 and p_2. All distortion coefficients and principal point $(c_x, c_y) \in \mathbb{R}^2$ are obtained from the camera calibration method from [13]. This non-linear mapping imposed by the extended camera model $p = \xi(x', y')$ can be numerically inverted. Thus, it is possible to determine for each point $p \in \mathbb{R}^2$ (or pixel location $\mathbf{x} \in \Omega$) its corresponding undistorted location $(x', y') = \xi^{-1}(\mathbf{x})$. Notice that the resulting undistorted location is a point whose coordinates are real value coordinates $(x', y') \in \mathbb{R}^2$. Hence, the estimation of the intensity value of an undistorted image requires an interpolation method which is based on the discrete pixel locations in Ω (see Eq. 4.1). For approaches based solely in point features, this can be done selectively in those regions in order to avoid unnecessary workload. However, in the visual perception for model coupling it is necessary to acquire complete undistorted images in order to extract geometric visual-features distributed across large regions of the images. For example, geometric edges and homogeneous segments. Based on these observations, the computation of the undistorted location (x', y') for each distorted location $\mathbf{x}$ is realized off-line and stored in a so-called undistortion map (similar to [12, 14]). In addition, since the bicubic interpolation [4] is selected due to its high acutance properties, the weighting coefficients used to estimate the intensity of each pixel position are also computed off-line and stored in the undistorion map. This dynamic programming approach improves the time performance of the high acutance undistorion process.

Another benefit of this process is the complete use of the acquired HDR distorted image. This means, when the undistorted locations (x', y') are outside the dimension of the image (see Eq. 4.1), a larger image is generated and the offset coordinates are properly mapped. Thus, the resulting undistorted image contains the complete field of view of the camera in a undistorted manner (see Fig. 5.4).

[3]Minimum object distance is the shortest distance between the lens and the subject.

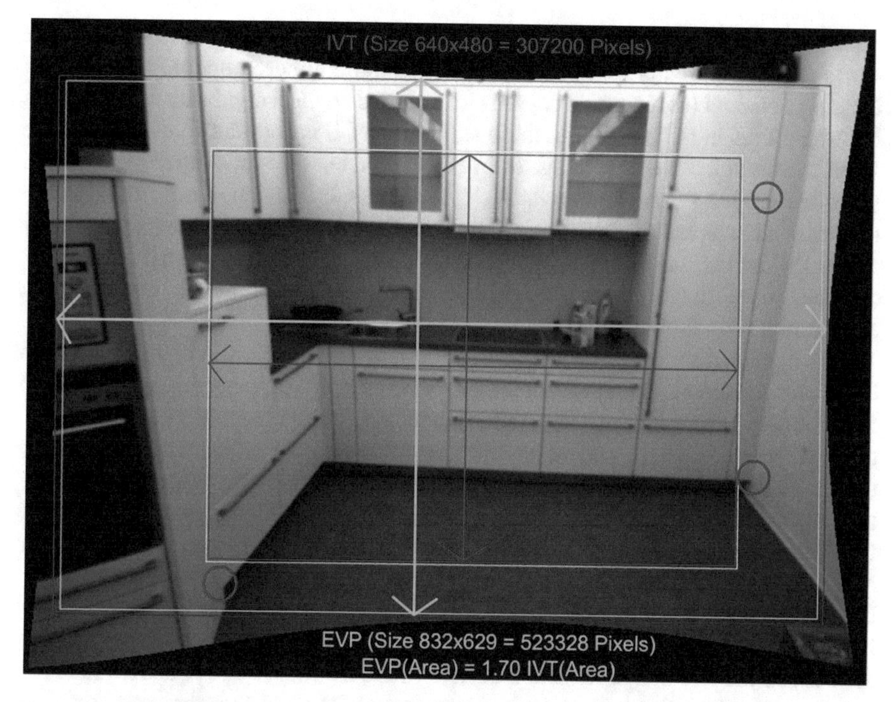

Fig. 5.4 The high acutance lens undistortion process applies bicubic interpolation [4] based on the radial and tangential distortion model [13]. The whole field of view is attained by an additional coordinate mapping in a larger image. The size of the resulting image depends on the lenses calibration. In this particular example, the camera [15] with a VGA (640 × 480 pixels) resolution coupled with a $f = 4$ mm lens produces an undistorted image containing more than 70% additional pixels than the distorted VGA source image. This is of particular importance when the robot is close to objects during grasping and manipulation. The whole undistorted field of view allows the maximal exploitation of the acquired HDR

5.5 Virtual Increase of Spatial Resolution

The spatial resolution of an image is the result of the lens properties (focal distance, minimum object distance, optical and chromatic aberration, etc.) and the image sensor resolution (pixel size and total pixels). In the best case, a high quality lens used with a high resolution camera will produce adequate images for visual perception. However, this straightforward approach is not always plausible in humanoid robots due to various embodiment restrictions (see Sect. 4.2). Additionally, images with high spatial resolution require more time to be transfered from the camera to the host CPU. This limits the frames per second FPS for the so-called in-the-loop visual tasks such as visual servoing and real-time tracking. High resolution images also require considerable computational resources to be processed. The increase of resolution implies a quadratic increment of the total amount of pixels. Therefore, a trade-off between image resolution and computational power has to be accomplished.

In order to increase the spatial resolution of a dedicated robot vision system there exists various approaches. These approaches require either structural similarity of the image content (as repetitive patterns in [16]) or statistical regularity with previous training set (as in [17]). Without these assumptions, it is not plausible to extend the spatial resolution of a single image in an efficient manner. Thus, an attempt to go into this direction is a research topic itself. Nevertheless, by reflecting upon the discrete operators such as the discrete convolution (in Sect. 5.6) or the 8-pixel connectivity within the dual non-maximal suppression (in Sect. 5.6.1) and based on experimental evidence, the following approach virtually increases the spatial resolution without any assumptions about image content.

From the theoretical point of view, the interpolation of an image produces no information increase or improvement. However, by proper image interpolation and kernel bandwidth selection, it is possible to obtain finer saliencies and more representative response functions during the filtering of the image (see Fig. 5.5). Practically, this emulates an increase of spatial resolution. This occurs during spatial convolution due to the reduced manifold support that the discrete filtering kernels (discrete masks) cover when directly applied on the source images. In particular, the finer response of the filtering kernels is notable at the corners or interstices between parallel structured elements like doors and drawers. This virtual resolution improvement has various advantages. For example, the edges found along the observed object are better extracted and represented because there are more edge pixels available to determine the curvature and (in general) the shape of the edge. Furthermore, because the edge junctions are difficult to process for most filtering methods based on linear filtering [2], interpolating the source image enables the robot application of Gestalt methods for junction extraction (see Sect. 5.6.1).

5.6 Saliency Extraction

Oriented Saliency Extraction: The receptive radiance saliency $\Gamma : \mathbb{N}^2 \mapsto \mathbb{R}^2$ is computed by the composed Gabor kernel (see Fig. 5.1). It is formally expressed as

$$\Psi(\mathbf{x}, \alpha, \gamma_1, \gamma_2, \lambda) = \underbrace{\mathbf{exp}\left[-\frac{1}{2}\mathbf{x}^T \Sigma^{-1}\mathbf{x}\right]}_{\text{Mahalanobis distance}} \underbrace{\sin\left[\pi \frac{U_{\langle\alpha\rangle} \cdot \mathbf{x}}{\lambda}\right]}_{\text{secondary axis distance}}. \tag{5.3}$$

In contrast to [5], this formulation provides the means to conveniently control the impulse response function by:

- **Kernel orientation**: In Eq. 5.3, the principal orthonormal semi-axes

$$U_{\langle\alpha\rangle} = [\cos\alpha \;\; \sin\alpha]^T, \tag{5.4}$$

and

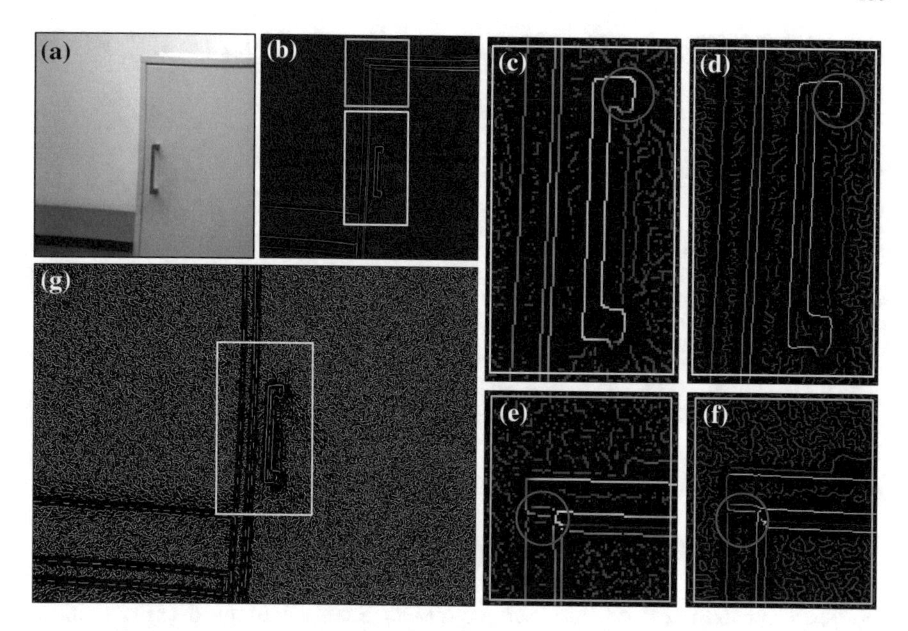

Fig. 5.5 The virtual spatial resolution increase is noticeable at the results of the processing pipeline. The improvements are detectable at corners or interstices between parallel structured elements like doors and drawers. **a** Source image. **b** Results of edge extraction (see Sect. 5.6). The green and yellow rectangles show the discussed effects. **c** Zoom of the resulting edge extraction from the source image. **d** The result of edge extraction by interpolating the image for increase of spatial resolution. At the magenta circles is possible to see the edge line segment. This results are no possible using the original source image due to the resulting spatial limitations. **e** Results from the original image show the lost of connectivity at this particular interstice of the fridge. **f** The improved results. **g** The difference between the source image and its improved version

$$V_{\langle\alpha\rangle} = [-\sin\alpha \;\; \cos\alpha]^T \,, \tag{5.5}$$

allow the saliency extraction in the α-target direction.

- **Kernel support**: Is shaped by the Mahalanobis distance adjusted by the ratio $\gamma_1 : \gamma_2$ in Eq. 5.3. It is represented by the covariance matrix $\Sigma = diag[\gamma_1 \; \gamma_2] \left[U_{\langle\alpha\rangle} \; V_{\langle\alpha\rangle}\right]$, this ratio adjusts the trade-off between smoothing and extraction.

The computation of the receptive radiance saliency Γ requires $k \geq 1$ band-pairs (α, q) of orientation-complementary kernels (see Fig. 5.6)

$$\Gamma(\mathbf{x}) = \sum_{p=1}^{k} \sum_{q=0}^{1} U_{\langle p\pi/2k\rangle} \left\{ \Phi(\mathbf{x}) \circledast \Psi(\mathbf{x}, p\pi/2k + \frac{q\pi}{2}, \gamma_1, \gamma_2, \lambda) \right\}, \tag{5.6}$$

where $\circledast$ denotes the discrete convolution operator (see Fig. 5.7).

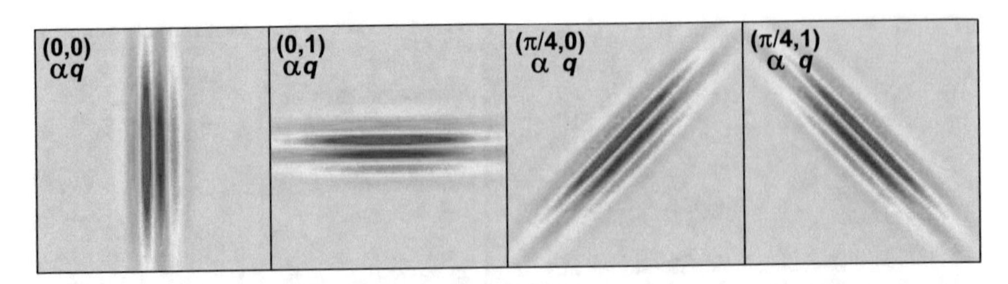

Fig. 5.6 The Gabor kernel set $\Psi(\mathbf{x}, \alpha + \frac{q\pi}{2}, \gamma_1, \gamma_2, \lambda),\ q \in \{0, 1\}$ is composed by orientation-complementary kernel band-pairs

5.6.1 Edge Extraction and Representation

Dual Non-maximal Suppression: Based on the saliency orientation

$$\widehat{\Gamma}(\mathbf{x}) = \Gamma(\mathbf{x}) \cdot \|\Gamma(\mathbf{x})\|^{-1},$$

the pixel location $\mathbf{x} \in \mathbb{N}^2$ has two neighbors $\tilde{\mathbf{x}}_{\pm} \in \mathbb{N}^2$ along $\widehat{\Gamma}(\mathbf{x})$ within its 8-neighborhood,

$$\tilde{\mathbf{x}}_{\pm} = \mathbf{x} \pm \left[\eta\left(\widehat{\Gamma}(\mathbf{x}), \widehat{i}\right), \eta\left(\widehat{\Gamma}(\mathbf{x}), \widehat{j}\right)\right]^T, \tag{5.7}$$

where the function $\eta : (\mathbb{R}^2, \mathbb{R}^2) \mapsto \{-1, 0, 1\}$ determines the discrete step along the unitary axes $\widehat{i}$ and $\widehat{j}$ as

$$\eta(a, b) = \begin{cases} 1, & \text{if} \quad \left(\widehat{a} \cdot \widehat{b} > cos(\frac{3\pi}{8})\right) \\ -1, & \text{else if} \left(\widehat{a} \cdot \widehat{b} < cos(\frac{7\pi}{8})\right) \\ 0, & \text{Otherwise.} \end{cases} \tag{5.8}$$

In contrast to other edge detectors, the dual non-maximal suppression $\Upsilon : \mathbb{N}^2 \mapsto \{0, 1\}$ selects pixels close to the continuous edge by considering not only the saliency norm $\|\Gamma(\mathbf{x})\|$ but also the saliency coherency μ (see Fig. 5.7d) as

$$\Upsilon(\mathbf{x}) = \begin{cases} 1, & \text{if} \left(\|\Gamma(\mathbf{x})\| > \|\Gamma(\tilde{\mathbf{x}}_+)\| \right) \wedge \left(|\widehat{\Gamma}(\mathbf{x}) \cdot \widehat{\Gamma}(\tilde{\mathbf{x}}_+)| > \mu \right) \wedge \\ & \quad \left(\|\Gamma(\mathbf{x})\| > \|\Gamma(\tilde{\mathbf{x}}_-)\| \right) \wedge \left(|\widehat{\Gamma}(\mathbf{x}) \cdot \widehat{\Gamma}(\tilde{\mathbf{x}}_-)| > \mu \right) \\ 0, & \text{otherwise.} \end{cases} \tag{5.9}$$

The resulting edge expose the expected minimal response (as in [18]). However, the Gabor energy [5] and the dual non-maximal suppression produce better edge maps than those resulting from the standard method [18] (see Fig. 5.8).

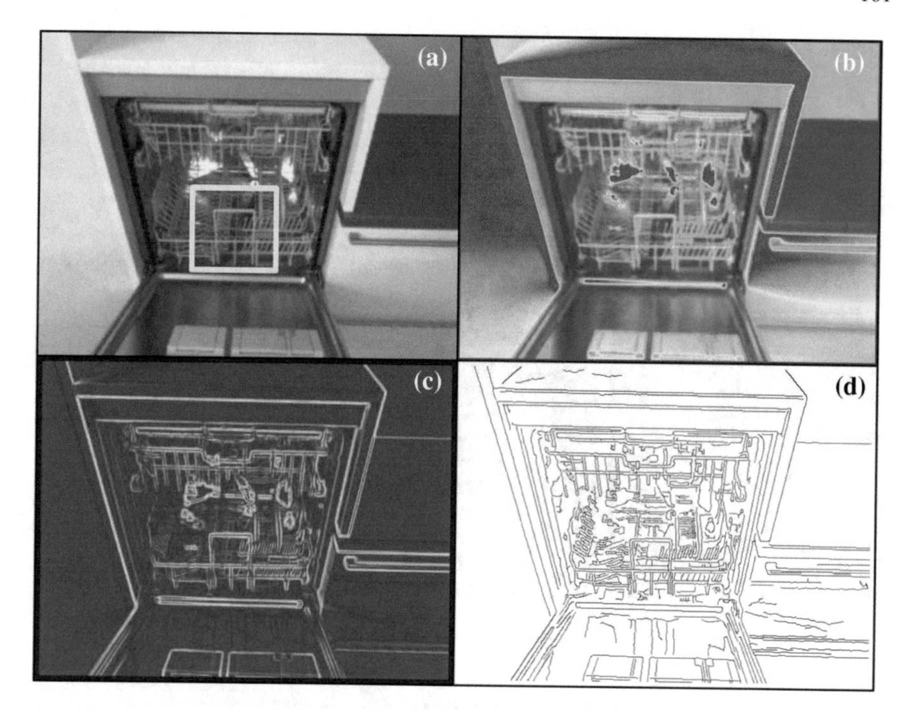

Fig. 5.7 Edge extraction process. **a** Wide intra-scene radiance range with complex structures and metallic surfaces. Notice the selected yellow window used in Fig. 5.8. **b** The HDR image Φ. **c** Receptive radiance saliency $\|\Gamma\|$ from Eq. 5.6. **d** Edge pixel map Υ from Eq. 5.9. The coherency factor $\mu = cos(\pi/8)$ partially removes phantom edges from the map and is tightly related to the spatial resolution per pixel (see Sect. 5.5)

Edge Subpixel Optimization: In order to properly extract the edge-graphs, it is necessary to optimize the edge-pixels to their subpixel position. Every pixel $\mathbf{x}$ on the binary map $\Upsilon(\mathbf{x})$ is located at most $\sqrt{2}$ units away from the local maximum edge on $\|\Gamma(\mathbf{x})\|$. In order to reach this maximum, the discrete location $\mathbf{x}$ is refined to its subpixel counterpart $\dot{\mathbf{x}} \in \mathbb{R}^2$ by gradient ascent along the radiance saliency direction using the bicubic convolution interpolation Λ (from [4], see Fig. 5.9a). The refinement is subject to

$$\dot{\mathbf{x}} = \Lambda[\mathbf{x}, \widehat{\Gamma}(\mathbf{x})] \Rightarrow \langle|\Gamma(\dot{\mathbf{x}})| \geq |\Gamma(\mathbf{x})|\rangle \wedge \langle|\dot{\mathbf{x}} - \mathbf{x}| < 1\rangle. \tag{5.10}$$

After this refinement, it is possible to exploit the discrete pixel grid for a graph expansion while establishing the node links according to the subpixel distances. In this manner, the structure of the radiance edge is correctly captured by the graph incidence with high reliability.

Edge-Graph Extraction: The graph extraction proceeds analog to the $\mathbf{A}^*$ algorithm [19] with a region growing approach. This means, the edge-graphs G_k are extracted

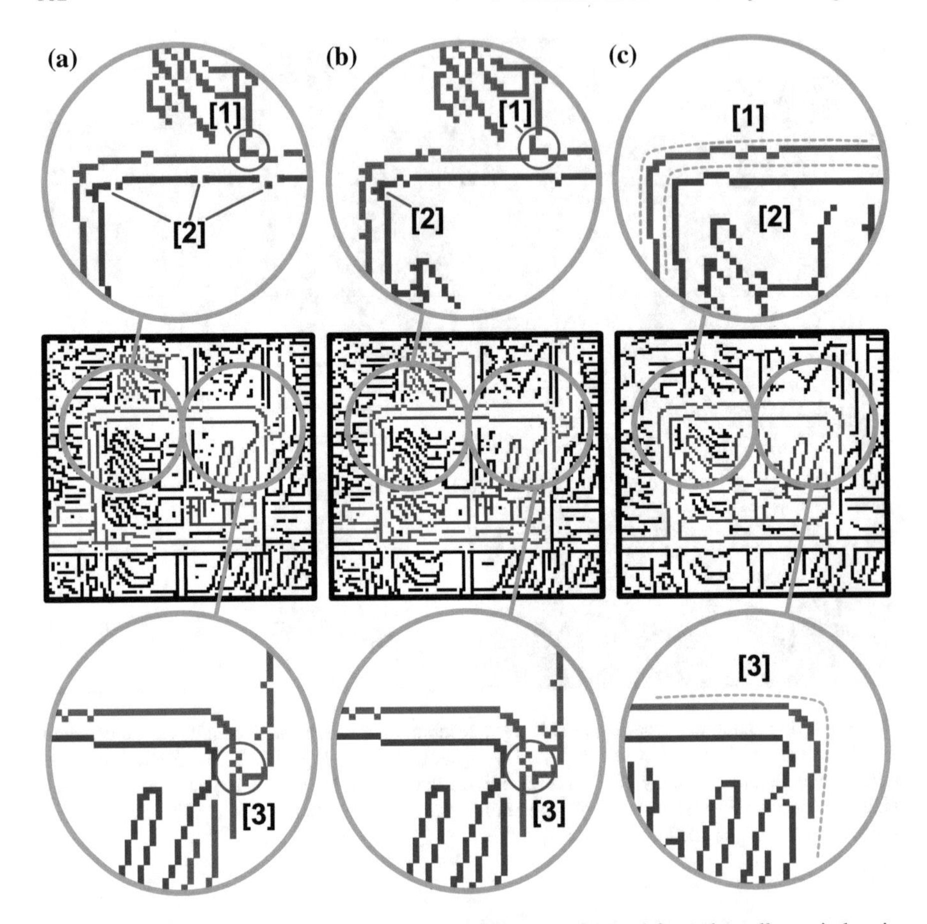

Fig. 5.8 Comparative edge-extraction. These results were obtained from the yellow window in Fig. 5.7a. The Canny filter applied with maximal sensitivity (the lower and the higher threshold set to 0, namely pure non-maximal suppression). **a** Column results from the implementation in IVT [12]. **b** Column results from the implementation in OpenCV [14]. **c** Edge resulting from the proposed method in this work supports the greater quality of the attained edges. Red edges obtained at the marked locations a-1, a-3, b-1 and b-3 present connectivity to residual clutter which corrupts the underlying geometric structure. Besides, the broken segments in a-2 forbid the primitive detection. In b-2, the fast but deficient Canny filter poorly reflects the inherent edge structure

by a heuristic recursive expansion. The extraction (traversal priority) stack is sorted by the subpixel distances $\left\|\dot{\mathbf{x}}_i - \dot{\mathbf{x}}_j\right\|$ (see Fig. 5.9b).

Edge-Graph Characterization: The edge-graphs G_k are acyclic linked collections of subpixel nodes, namely edge-trees. They expand along the radiance edges revealing their distinctive structural composition. In order to properly characterize this composition, a preprocess is conducted.

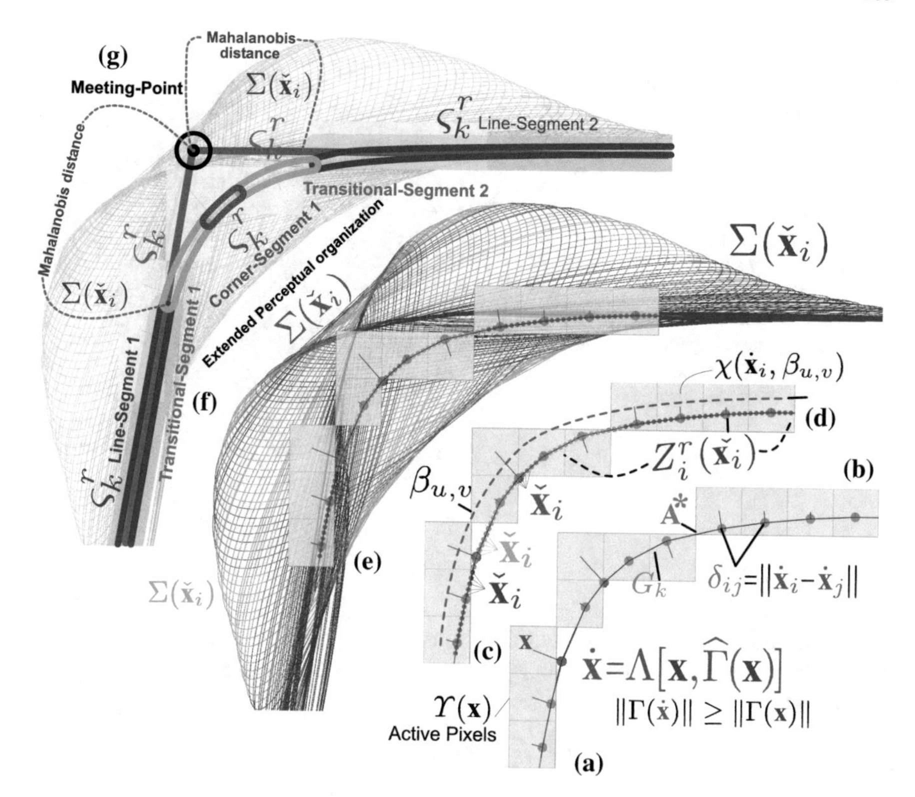

Fig. 5.9 Different eccentricity edge-graph processing stages. **a** Subpixel optimization (from Eq. 5.10). **b** Graph extraction. **c** Outspreading process (see Eq. 5.11). **d** r-Zone node set. **e** Eccentricity ellipses from the Eigenvalues and Eigenvectors of the covariance matrix (in Eq. 5.14). **f** Geometric primitive segmentation. **g** Evidence-based grouping transform

- **Pruning**: It improves the representativeness of the linkage by eliminating superfluous leaf nodes of the trees. This is done by a tree-depth analysis. In this process redundant leaf nodes are removed by straightforward selection base on depth index.
- **Outspreading**: It enhances the cohesiveness of the graph linkage with the Euclidean distances of the nodes. The outspreading is performed by uniform arc-length arrangement $\chi : \mathbb{R}^2 \mapsto \mathbb{R}^2$ of the $\dot{\mathbf{x}}_i$ nodes contained within the tree branch paths $\beta_{u,v}$ (see Fig. 5.9c),

$$\check{\mathbf{x}}_i = \chi(\dot{\mathbf{x}}_i, \beta_{u,v}). \tag{5.11}$$

During this arrangement, the branch-paths are treated as parametric curves where the supporting points are subpixel node locations $\dot{\mathbf{x}}_i$. In this way, the implicit parametric curve of the edge branches remains unaffected, while the graph and Euclidean distance are cohesively correlated. The resulting cohesively combined statistical distribution is the key to characterize the nodes $\check{\mathbf{x}}_i$ by the proposed eccentricity model ξ of their graph neighborhood. This neighborhood is called the *r-zone* and it is composed by the node subset

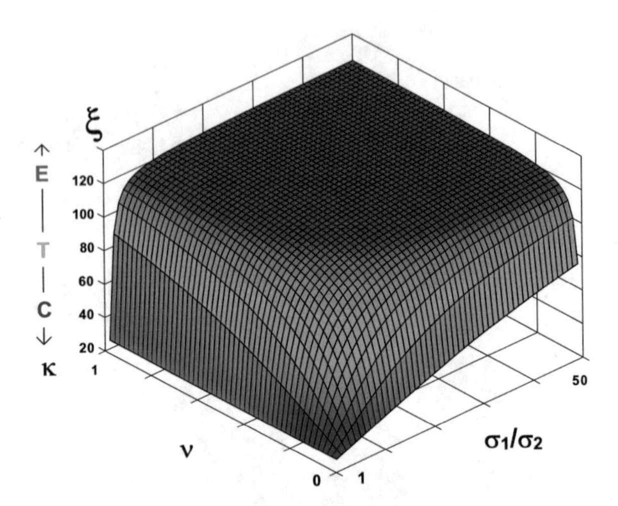

Fig. 5.10 The characterization eccentricity model $\xi(\kappa, \nu, \frac{\sigma_1}{\sigma_2})$. Notice the segmentation into three types. The first category corresponds to the elongated neighborhoods E with high eccentricity level. The second category are the transitional neighborhoods T with middle eccentricity level. Finally the compact neighborhoods C with low eccentricity level (see also Table 5.1)

$$Z_i^r(\check{\mathbf{x}}_i) := \{\check{\mathbf{x}}_j\}. \tag{5.12}$$

These are the nodes reachable within a maximal r graph-distance (see Fig. 5.9d). The r-zone cardinality is denoted as $|Z_i^r|$ (see Fig. 5.9e). The eccentricity model ξ is expressed as

$$\xi(\check{\mathbf{x}}_i) = \kappa \left[\frac{2}{\pi} \arctan\left(\nu \frac{\sigma_1}{\sigma_2} \right) - \frac{1}{2} \right], \tag{5.13}$$

where σ_1 and σ_2 are the Eigenvalues of the $\check{\mathbf{x}}_i$-centered covariance matrix,

$$\Sigma(\check{\mathbf{x}}_i) = \frac{1}{|Z_i^r|} \sum_{j=1}^{|Z_i^r|} (\check{\mathbf{x}}_j - \check{\mathbf{x}}_i)(\check{\mathbf{x}}_j - \check{\mathbf{x}}_i)^T. \tag{5.14}$$

In Eq. 5.13, the saturation factor ν shifts the clipping Eigenratio, whereas the normalization, offset and amplitude factors conveniently shape the eccentricity response function ξ (see Fig. 5.10) in order to suitably distinguish the $k := \{\mathbf{C}, \mathbf{T}, \mathbf{E}\}$ eccentricity types (see Table 5.1).

Edge-Graph Segmentation: Geometric-primitives emerge by grouping nodes according to their k-type. These cliques $\varsigma_k^r := \{\check{\mathbf{x}}_i\}$ constitute nodes in a geometric-primitives graph G_i^r (see Fig. 5.9f).

Evidence-based Perceptual Organization: In contrast to [20, 21], *proximity* and *continuity* among geometric-primitives are not detected exclusively based on the length, position, orientation and distance between endpoints. The improved detection incorporates structural likelihood by the Mahalanobis distance induced between the edge-node combined statistical distributions $\Sigma(\check{\mathbf{x}}_i)$ (from Eq. 5.14) of the periphery of the clique and the intersection point (see Fig. 5.9g). This (morphologic and evidence-

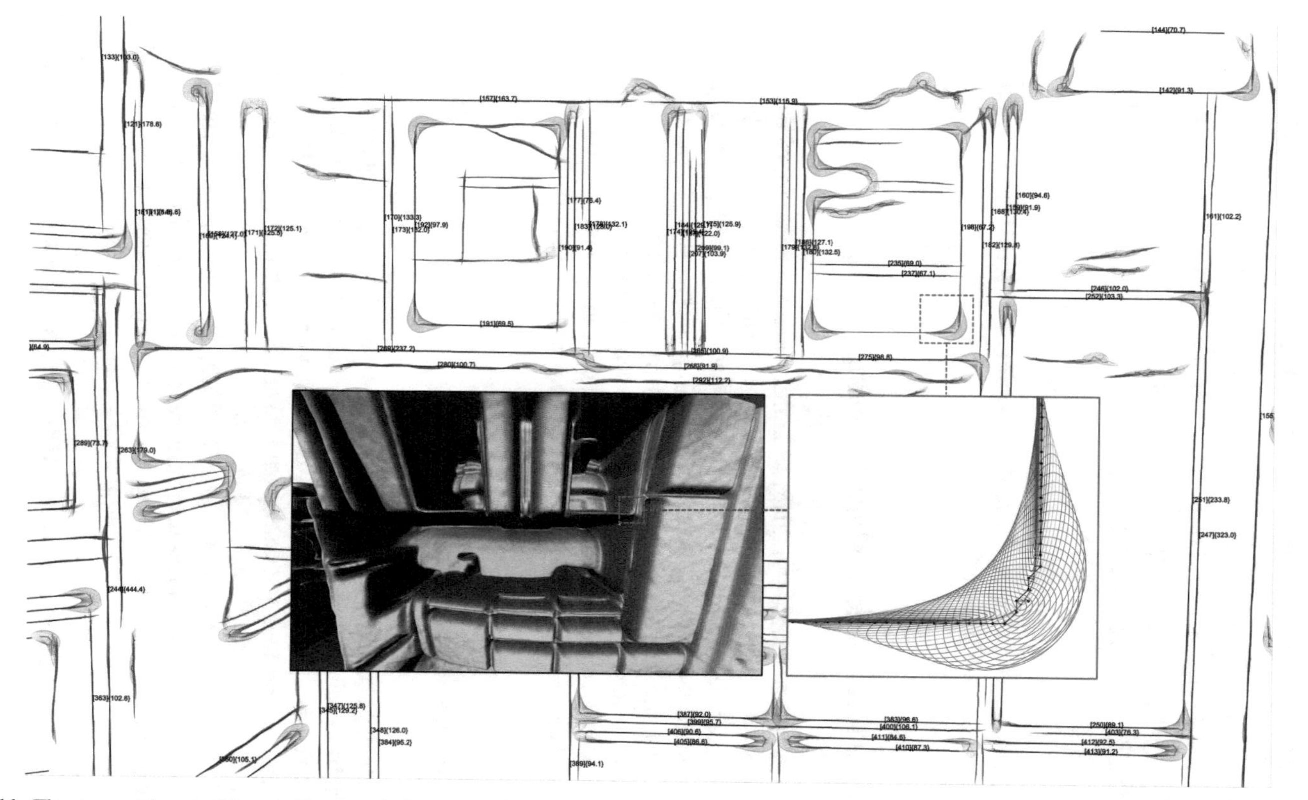

Fig. 5.11 The geometric-primitives (with extended perceptual organization) are the result of the processing pipeline (see Fig. 4.28). The zoom shows the subpixel edges with eccentricity ξ_i (see Table 5.1)

Table 5.1 Characterization of eccentricity model ξ for graph saliencies. This table presents the eccentricity levels used to segment the edges according to their neighborhood characterization. This categorization provides the means to segment edge paths according to their shape into lines, curves and junctions

k-Type	ξ -Range	ξ -Level	$\sigma_1 : \sigma_2$	Segment	Z_i^r -Subtree
E - *Elongated*	$[\xi_H, \infty)$	High	$\sigma_1 \gg \sigma_2$	Line	Paths
T - *Transitional*	$[\xi_L, \xi_H)$	Medium	$\sigma_1 > \sigma_2$	Curve	Paths
C - *Compact*	$[1, \xi_L)$	Low	$\sigma_1 \approx \sigma_2$	Junction	Leafs/Splitters

based) grouping transforms the geometric-primitives from ς_k^r nodes to consistently matchable geometric percepts.

5.7 3D Geometric Feature Matching

The extracted geometric-primitives from the calibrated stereo-rig of the humanoid robot are used to calculate their 3D pose [11, 12]. The resulting Euclidean length and angles of the percepts are reasonably pose-invariant. These geometric-primitives are organized and grouped into a 3D spatial graph representation. This visual-feature representation allows the application of the interpretation tree from [22]. The hypothesis are constructed by pair of visually perceived line segments and the probabilistic indexing model edges from Sect. 3.4.5. The application of this approach is analyzed and evaluated in terms of performance and precision in the Sect. 5.8.

5.8 Experimental Evaluation

The saliency extraction based on HDR images for edge-graph requires 1.68s with 2 kernel band-pairs. The graph extraction and eccentricity characterization take 1.46s with a standard deviation of 3.52s depending on the image content. Finally, the geometric regressions and perceptual organization estimate the line segments and meeting points in 25–50 ms (see Fig. 5.9).

In the presented experiments (see Fig. 5.12), the matching method used to recognize the grasping-target in the dishwasher exploits the stereo Euclidean metric to find a line-segment node with a known length (116mm), namely the grasping segment between two incident line-segment nodes. The complete single threaded execution time of 10–12s can be reduced by at least two thirds. Notice that there is an upper performance boundary by the delay of the camera response to exposure commands.

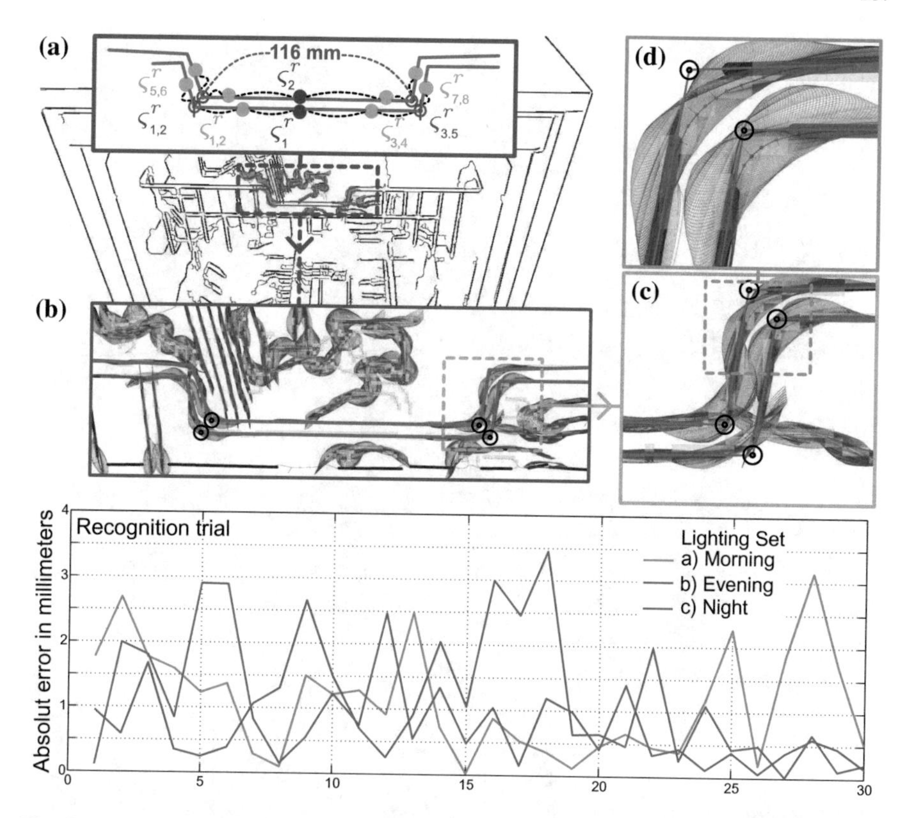

Fig. 5.12 On the upper plot, the recognition of a dishwasher basket is a perception challenge for the complexity of the overlapping structures and the wide intra-scene radiance. **a** Scalable vector graphics output of the recognition component. The selected region shows the grasping target region to be recognized. The upper schematic shows the geometric graph to be matched. **b** First zoom. **c** Second zoom. **d** At the third zoom, it is possible to see the eccentricity types in red, green and blue (covariance ellipses from Eq. 5.14). The subpixel graph configuration along the curve of the metal wire of the basket shows the simultaneous extraction and segmentation of geometric-primitives. On the lower plot, the precision is shown by the absolute error plots. The ground truth is the 116 mm length of the physically measured segment. **a** The plot shows the precision attained with artificial lighting inclusive natural morning illumination incoming from the windows. **b** Plot with artificial lighting inclusive natural evening illumination. **c** Only artificial lighting (see Table 5.2)

In the evaluation, the accomplished recognition rate of 93.04% can be improved by using more advanced matching methods [20, 23] (Fig. 5.11). When computing the length of the grasping segment, a maximal absolute error of 3.45 mm was obtained and no detrimental performance was measured by lighting variations (see Figs. 5.12, 5.13 and Table 5.2).

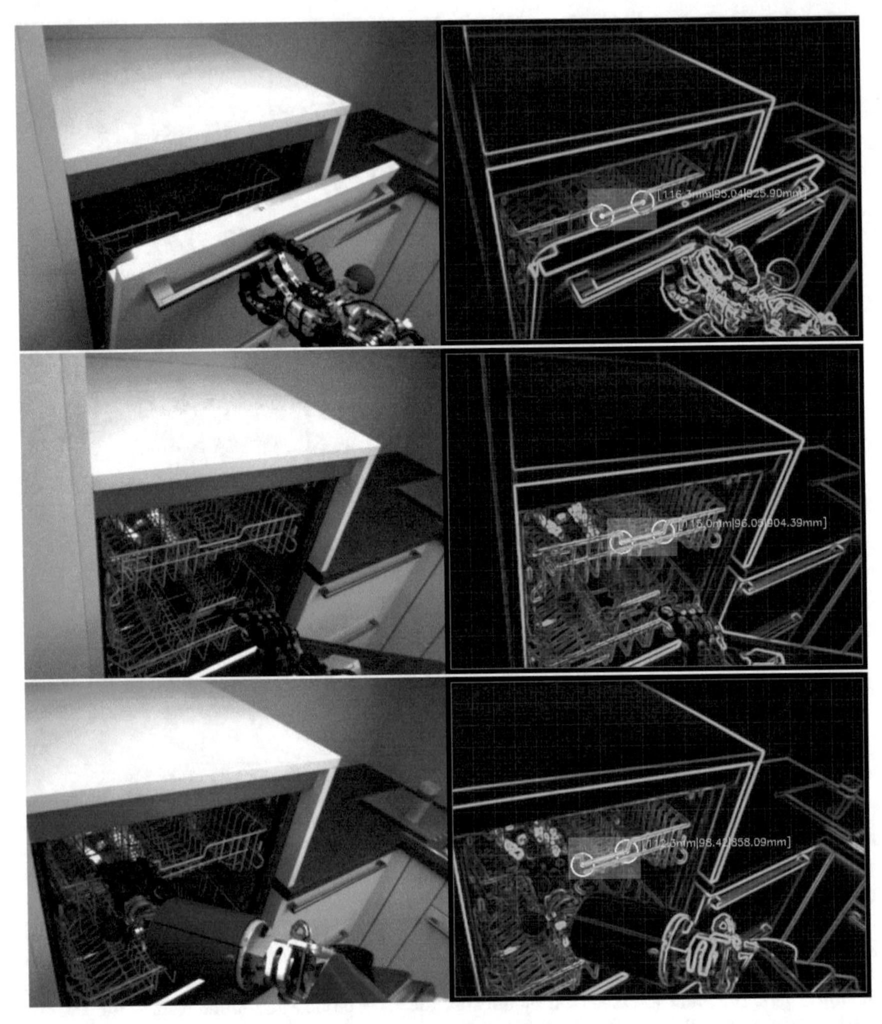

Fig. 5.13 Visual recognition and pose estimation of the dishwasher basket. This three stacked images show on the left column the input image of the left camera of the humanoid robot ARMAR-IIIa while opening the dishwasher (see motion planning and control in [24]). On the right column, the visualization of the saliency function with the marked recognized element is visualized. This illustrates the visual recognition for grasping and manipulation based on the proposed method. The qualitative evaluation is presented in Table 5.2

5.9 Discussion

The presented method has no parameters which depend on the image content. The content independent parameters discussed here have only well-bounded impact on the performance of the execution (Fig. 5.14). In the experiments, the exhaustive geometric-primitives search took at most 7–10s using a Intel-CPU @ 1.8 GHz with

Table 5.2 The precision performance with different illumination setups (see Fig. 5.12). Notice the maximal absolute error at 1064.23 mm between the camera and the recognized object is 3.45 mm. This precision corresponds to the physical limitations of the actuators in mobile robots. Hence, the method is appropriated for physical interaction in terms of manipulations and grasping

Set	Mean length	Std. deviation	Max-abs error	Distance
Morning	116.91	1.07	3.15	1032.62
Evening	116.31	1.18	2.9	1043.38
Night	116.57	1.28	3.45	1064.23

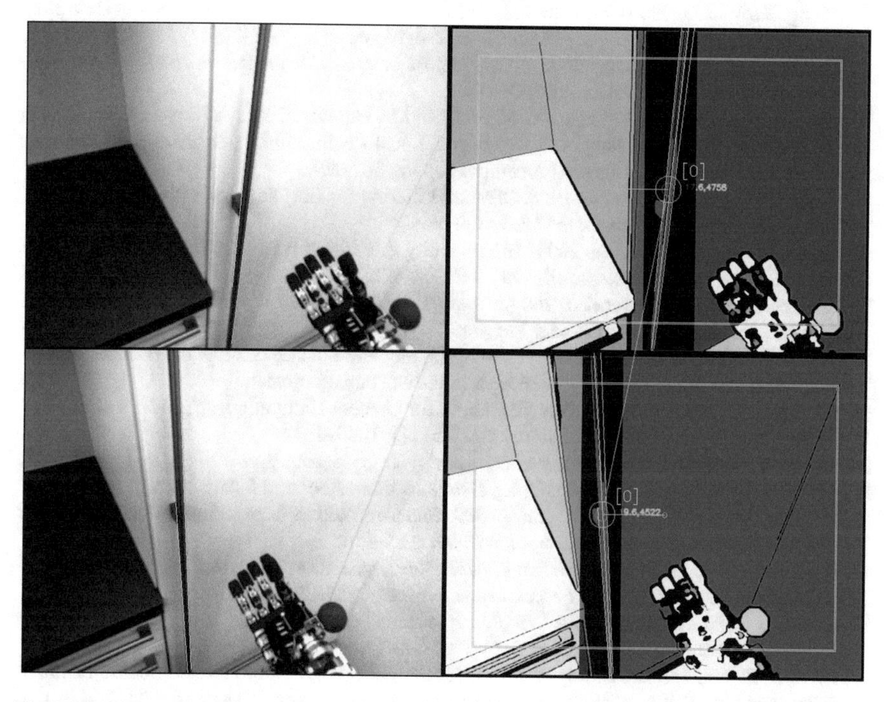

Fig. 5.14 Visual recognition and pose estimation of the fridge handle. The right column shows the input images from the left camera at upper row and the right camera on the lower row. The right column shows the results of the segmentation based on homogeneity and the edge analysis for the estimation of the location of the handle

no code optimization. Technical improvements such as parallelization using a GPU and faster interfaces to HDR-enabled cameras will improve the performance to real-time execution.

References

1. Ayache, N. 1991. *Artificial Vision for Mobile Robots*. MIT Press. ISBN 978-0262011242.
2. Papari, G., and N. Petkov. 2011. Edge and Line Oriented Contour Detection: State of the Art. *Image and Vision Computing* 29 (3): 79–103.
3. Ramanath, R., and W. Snyder. 2003. Adaptive Demosaicking. *Journal of Electronic Imaging* 12: 633–642.
4. Keys, R. 1981. Cubic Convolution Interpolation for Digital Image Processing. *IEEE Transactions on Acoustics, Speech and Signal Processing* 29 (6): 1153–1160.
5. Grigorescu, C., N. Petkov, and M. Westenberg. 2003. Contour Detection based on Nonclassical Receptive Field Inhibition. *IEEE Transactions on Image Processing* 12 (7): 729–739.
6. Wang, X., W. Lin, and P. Xue. 2005. Demosaicing with Improved Edge Direction Detection. *In IEEE International Symposium on Circuits and Systems* 3: 2048–2051.
7. Hirakawa, K., and T. Parks. 2005. Joint Demosaicing and Denoising. *In IEEE International Conference on Image Processing* 3: 309–312.
8. Moriya, S., J. Makita, T. Kuno, N. Matoba, and H. Sugiura. 2006. Advanced Demosaicing Methods based on the Changes of Colors in a Local Region. In *International Conference on Consumer Electronics, Digest of Technical Papers*, 307–308.
9. Tomasi, C., and R. Manduchi. 1998. Bilateral Filtering for Gray and Color Images. In *International Conference on Computer Vision*, 839–846.
10. Lukin, A., and D. Kubasov. 2004. High-Quality Algorithm for Bayer Pattern Interpolation. *Programming and Computer Software* 30 (6): 347–358.
11. Hartley, R., and A. Zisserman. 2004. *Multiple View Geometry in Computer Vision*, 2nd ed. Cambridge: Cambridge University Press. ISBN 0521540518.
12. KIT. 2011. Karlsruhe Institute of Technology, Computer Science Faculty, Institute for Anthropomatics, The Integrating Vision Toolkit. http://ivt.sourceforge.net.
13. Zhang, Z. 2000. A Flexible New Technique for Camera Calibration. *IEEE Transactions on Pattern Analysis and Machine Intelligence* 22 (11): 1330–1334.
14. OpenCV. Open Source Computer Vision Library. http://opencv.org.
15. PTGrey. 2008. *Dragonfly Technical Reference Manual*. Accessed 5 Aug 2008.
16. Glasner, D., S. Bagon, and M. Irani. 2009. Super-resolution from a Single Image. In *IEEE International Conference on Computer Vision*, 349–356.
17. Park, S.C., M.K. Park, and M.G. Kang. 2003. Super-resolution Image Reconstruction: A Technical Overview. *IEEE Signal Processing Magazine* 20 (3): 21–36.
18. Canny, J. 1986. A Computational Approach to Edge Detection. *IEEE Transactions on Pattern Analysis and Machine Intelligence* 8 (6): 679–698.
19. Hart, P., N. Nilsson, and B. Raphael. 1968. A Formal Basis for the Heuristic Determination of Minimum Cost Paths. *IEEE Transactions on Systems Science and Cybernetics* 4 (2): 100–107.
20. Lowe, D. 1987. Three-Dimensional Object Recognition from Single Two-Dimensional Images. *Artificial Intelligence* 31: 355–395.
21. Sarkar, S., and K. Boyer. 1993. Perceptual Organization in Computer Vision: A Review and a Proposal for a Classificatory Structure. *IEEE Transactions on Systems, Man and Cybernetics* 23 (2): 382–399.
22. Grimson, W. 1990. *Object Recognition by Computer: The Role of Geometric Constraints*. Cambridge, MA, USA: MIT Press. ISBN 0262071304.
23. Wolfson, H., and I. Rigoutsos. 1997. Geometric Hashing: An Overview. *IEEE Computational Science Engineering* 4 (4): 10–21.
24. Vahrenkamp, N., T. Asfour, and R. Dillmann. 2012. Simultaneous Grasp and Motion Planning. *IEEE Robotics and Automation Magazine*, 43–57.

Chapter 6
Visual Uncertainty Model of Depth Estimation

The visual skills linking the physical world with the internal world representation of a humanoid robot are affected by internal and external uncertainty. These uncertainties are present at various levels ranging from noisy signals and calibration deviations of the physical system up to the limited granularity and mathematical approximations of the perception-planning-action cycle. This cumulated uncertainty limits the precision and efficiency of the visual perception. In order to overcome these limitations, the depth uncertainty has to be modeled in the skills of the humanoid robot. The model of perceptual uncertainty is important for the robustly realization of complex actions. Furthermore, the integration of the visual depth uncertainty model in task planning and execution enables better perceptual scalability by managing ambiguities and handling errors while simultaneously enables consistent sensor fusion. Due to the complexity of the inherent hardware-dependent uncertainties in humanoid robot perception, the visual uncertainty can be hardly modeled analytically. However, the uncertainty distribution can be conveniently learned with the help of supervised methods. In this approach, the role of the supervisor is to provide ground truth spatial measurements corresponding to the uncertain observations. In this chapter, a supervised learning method for inferring a novel model of visual depth uncertainty is presented. The acquisition of the model has been autonomously generated by the humanoid robot ARMAR-IIIb.

6.1 Uncertainty Modeling: Related Work

Humanoid robots have to perceive the real world in order to attain the vision model coupling (see Sect. 2.4). The perceptive system should quantitatively determine the observed properties of the attributes and relations in the world such as size, location and orientation. Additionally, the perceptive system should qualitatively solve complex tasks such as recognition, classification and interpretation. In the vision

D. I. González Aguirre, *Visual Perception for Humanoid Robots*,
Cognitive Systems Monographs 38, https://doi.org/10.1007/978-3-319-97841-3_6

model coupling, the quantitative perception provides the essential cues (length and depth) for complex perception tasks of object recognition with 6D-pose estimation. This visual quantitative perception relies on different cues depending on the sensing approach. Evidently for humanoid robots the most coherent and natural approach is the use of stereo vision to acquire quantitative information.

The uncertainty analysis and modeling for stereo vision has been an active research field in the last three decades. Considerable results have been achieved (see approaches from the pioneers [1–3] up to recent contributions [4–6]). Despite their significance, previous approaches rely on at least one fundamental and critical assumption:

- **Parametric distribution**: They consider an underlying parametric distribution of the perceptual deviations and usually model it by arbitrary fitting parametric distributions. This occurs either explicitly or quite subtly.
  - **Explicit parametric model**: This occurs when modeling the uncertainty as the weighted sum of normal distribution(s) without rigorous validation of the uncertainty distribution profile, namely a methodical analysis of the plausibility of a parametric distribution [2]. This rigorous modeling includes the analysis of the *systematic errors* and the *stochastic spreading* as functions of depth. These approaches also lack of the validation through deviation analysis between their estimated distribution and one accurate ground truth-based non-parametric model which tightly reflects the uncertainty nature of the visual perception.
  - **Implicit parametric model**: These assumptions are implicitly stated by a priori covariance propagation. This takes place assuming simplifications through analytical models satisfying strong mathematical constraints, usually depicted by linearization through truncated Taylor series [6].
- **Actuator uncertainty effects**: Previous approaches also ignore the actuator effects within the active visual perception and consider only one combination of image and feature extraction [3].
- **Ground truth reference**: These methods lack of independent and accurate source of measurements to widely (in contrast to [6] with several meters of depth), strictly and trustworthy validate their models within a deep range and a large amount of trials.

6.2 Supervised Uncertainty Learning

In this chapter, a novel uncertainty model of the visual depth perception is presented. In contrast to the discrete approaches (as in [5]), the following method is based on supervised uncertainty learning in continuous space. In these terms, the uncertainty model of stereoscopic depth perception is the inferring function Ψ which maps the visual depth (distance between the camera and the target object) δ to its ground truth depth γ in terms of the probability density function ζ, namely

$$\underbrace{\{\Psi\}}_{\text{ground truth model}} \quad : \quad \underbrace{\{\delta, \gamma \in \mathbb{R}\}}_{\text{perceptual and ground truth depths}} \quad \mapsto \quad \underbrace{\{\zeta(\delta, \gamma) : \mathbb{R}^2 \mapsto \mathbb{R}\}}_{\text{ground truth PDF}}.$$

The realization of this non-reductionist approach takes four compounded and interconnected elements (see Fig. 6.3):

- **Ground truth kinematic tree**: This tree connects the ground truth 6D-pose of the robot cameras and the visually recognized reference-rig(s). The external reference is partially attained with the marker-based system. Notice that the reference system can only partially provide the required information to build the kinematic tree. It occurs because the kinematic frame of the camera C cannot be straightforwardly obtained by placing markers at the humanoid robot head nor by attaching markers close to its eyeballs (see Fig. 6.7). In order to estimate the stereo camera kinematic frame C, a precise registration process must be conducted in order to link the visual recognition with the marker system (see Sect. 6.6).
- **Visual perception pipeline**: The visual perception consisting of stereo camera calibration, feature extraction and depth estimation methods to be analyzed in order to modeled into a uncertainty distribution (see Chap. 5).
- **Autonomous uncertainty sampling**: An autonomous process for collecting learning samples. This process generates a scanning plan, controls its execution on the humanoid robot using six DoFs, three DoFs of the platform and three DoFs of the neck. The scanning plan simultaneously coordinates the network interfaces with the marker based ground truth system in order to gather the learning samples. This autonomous sampling collects large sample sets necessary for soundly inferring the uncertainty model.
- **Learning and modeling methods**: The inferring method analyzes the acquired learning samples to obtain the uncertainty model function Ψ. Afterwards, the detailed analysis of the model unveils the systematic-errors and stochastic-spreading of the visual perception. This produces an outstanding uncertainty model representation in terms of usability and minimal computational complexity.

In Sect. 6.3, particular aspects of the depth perception are discussed. Subsequently, the Sect. 6.4 describes the modeling setup used for supervised learning. Afterwards (in Sect. 6.5), the visual recognition and 6D-pose estimation of the reference-rigs are introduced. Subsequently, the Sect. 6.6 describes the ground truth acquisition subsystem. Thereafter in Sect. 6.7, the integration of visual recognition and ground truth into the uncertainty model Ψ is presented including its representations. Finally, the Sect. 6.9 discusses the proposed model.

6.3 Visual Depth Perception Uncertainty

In order to sagaciously model and manage the uncertainty of a robot vision system, it is crucial to reflect upon the nature of the acquired uncertainties. There are two

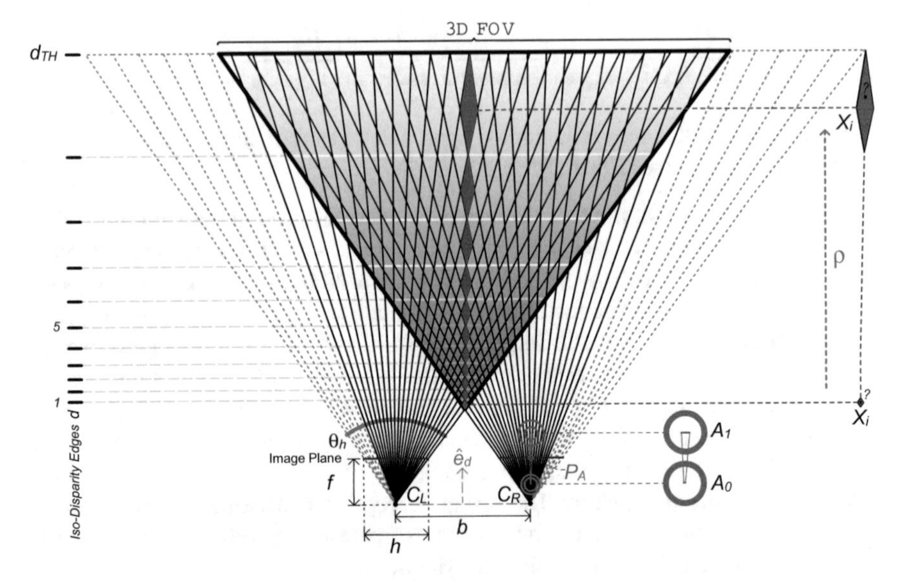

Fig. 6.1 The *image-to-space uncertainty* factors of an ideal stereo front-parallel configuration. The volume contained within the intersection of the frustum of two pixels (displayed in red) directly depends on its depth, base line length and focal distances of the stereo rig

remarkable categorical sources of uncertainty, namely the *image-to-space* and the *space-to-ego* uncertainties.

6.3.1 Image-to-Space Uncertainty

The image-to-space uncertainty is introduced at the visual recognition level and pose estimation processes. Its generation begins with the pixel precision limitations, such as noise, discretization, quantization, etc., terminating with error-limitations of the camera model and calibration, namely radial-tangential distortion and intrinsic parameters [8]. This uncertainty source is strongly correlated with the perceptual depth.

In detail, the uncertainty arises from the following superposed facts: First, considering only the monocular influence in each camera of the stereo rig. The surface patch A_i on the plane perpendicular to the optical axis of the camera (Fig. 6.1) which is imaged to a single pixel P_A grows linearly as function of the depth $A_i = 4\delta_i \tan\left(\frac{\theta_h}{2h}\right) \tan\left(\frac{\theta_v}{2v}\right)$, where θ_h and θ_v represent the horizontal and vertical angular apertures of the field of view, whereas h and v depict the width and height resolutions of the image (see Fig. 6.1).

Additionally, the stereo triangulation has an additional effect during the estimation of the 3D position of a matched pair of points. The depth δ_i affects the magnitude of the disparity d_i. Therefore, the precision of the pixel computations plays a central role. This occurs due to fact that points close to the stereo rig base line have wider disparities along the epipolar lines, meanwhile those points located beyond distance[1] $\delta_{Th} > fb$ have a very narrow disparity falling into the subpixel domain $d < 1$. This results in inaccurate depth calculations. These situations produce sparse distributions of the iso-disparity surfaces [9]. In other words, the subspace contained between these surface-strata grows as $d_i = fb/\delta_i$, where the focal distance f and the base line size $b = ||C_L - C_R||$ play relevant roles in the measurement precision.

Figure 6.1 shows the ideal front parallel case with iso-disparity edges delineating the subspaces contained between two discrete steps in the disparity d_i. In this manner, points contained within one of these subspaces produce the same discrete disparity when matching corresponding pixels. Hence, the location uncertainty muss be proportional to the space contained between iso-disparity surfaces. These two factors produce an uncertainty growing in an attenuated polynomial fashion. This behavior arises naturally and is reflected in the model obtained in Sect. 6.7.2.

6.3.2 *Space-to-Ego Uncertainty*

The space-to-ego uncertainty is introduced while relating the position of a 3D visual estimated point from the camera frame to the ego frame, namely at the kinematic mapping between cameras and frames of the humanoid robot (see Fig. 6.2a). This is caused by the physical and measurement inaccuracies of the head which are substantially amplified by projective effects.

The almost negligible errors in the encoders and mechanical joints of the head of the humanoid robot are amplified proportional to the depth distance δ_i between the ego center and the location of the points in space. Figure 6.2b shows this kinematic chain starting at the visual estimate point x_i. Subsequently, the transformation from the camera frame to the platform frame passing through the eyes and neck frames. These are time varying transformations during the execution of the scanning strategy.

The estimation of the 3D position by triangulation is subject to inaccuracies due to the lack of subpixel calculations [10] and because of the flaws in the extrinsic calibration (see [4]). The depth uncertainty is an intricate heterogeneous composition. Many attempts to partially model these effects have been done [1–6]. However, non of them has simultaneously regarded all the facts that a complete analytical uncertainty model has to consider inclusively the particular image content. In real applications robots are exposed to a wide variety of materials, lighting and operation conditions. Additionally, there are divers feature extraction methods which have been successfully and complementary applied in complex humanoid robots application (see [11, 12]). Thus, a complete analytical formulation is unfeasible. Still, the robot depth perception must incorporate an effective uncertainty model which directly and consistently reflects the nature of the visual perception. A plausible model of

[1] The depth δ_{Th} of a point in the scene produces a disparity smaller than a pixel.

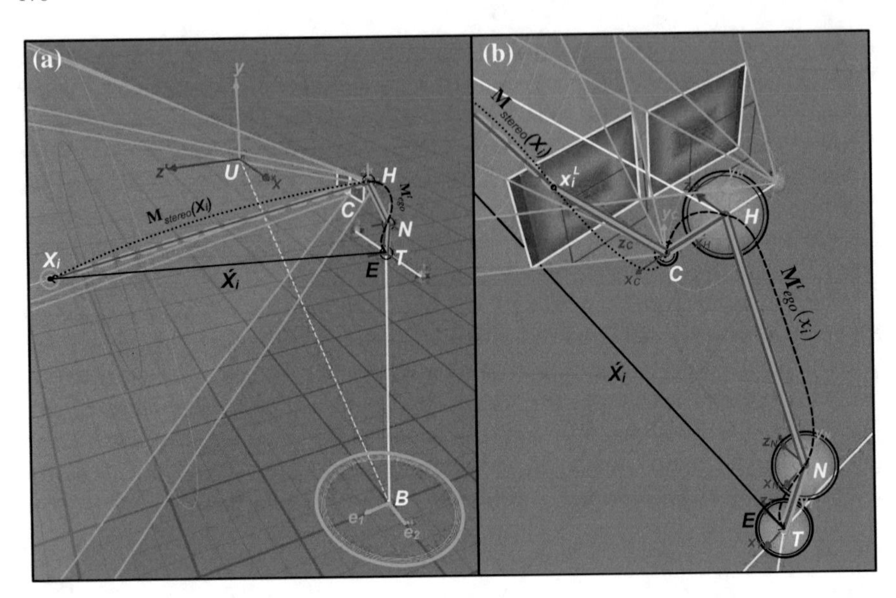

Fig. 6.2 The space-to-ego uncertainty is introduced by the mapping of visual estimated points from the camera frame to the ego frame. **a** Whole transformation from the visual space to the ego frame. **b** Partial transformation from the camera frame to the ego frame

such a complex composition is to represent the depth uncertainty Ψ in terms of the depth density distribution ζ, namely the PDF associating the perceptual depth δ and the ground truth depth γ. This holistic approach integrates all the uncertainty sources into a learnable compounded process providing a convenient description of the uncertainty distribution. Additionally, the proposed approach generates an optimal representation for pose estimation exploited in Chap. 7. In contrast to previous analytical approaches, the key concept in this work is to learn the non-parametric uncertainty model Ψ as a probability density function by kernel density estimation. The resulting non-parametric PDF Ψ is further analyzed by nonlinear regression in order to compare and verify its shape against the normal distributed uncertainty model.

6.4 Modeling Setup

A learning observation $S_t \in \mathbb{R}^2$ of the depth uncertainty probability function consists of the vision estimated depth $\delta \in \mathbb{R}$ from stereo images and the ground truth depth $\gamma \in \mathbb{R}$ attained by the marker system. In order to collect these observations, the following setup (in Fig. 6.3) is proposed:

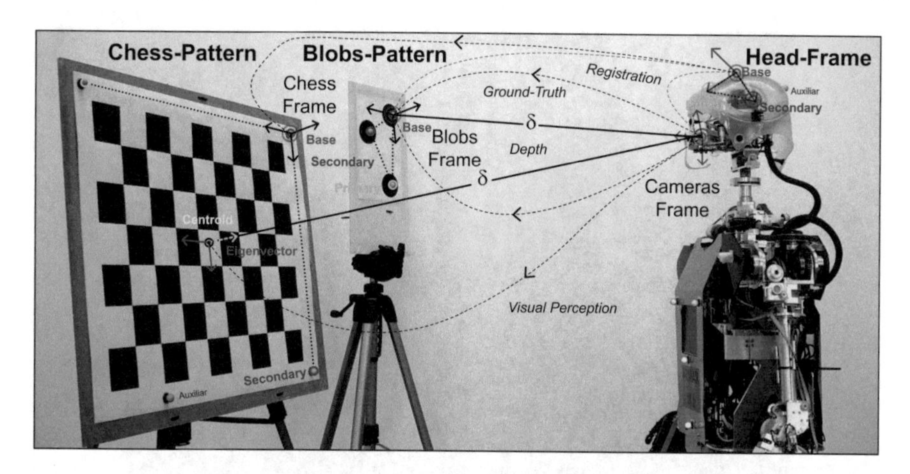

Fig. 6.3 Supervised learning of the visual depth uncertainty model consists four compounded elements: (i) Three markers on the humanoid robot head. (ii) The 2D blob-pattern reference-rig. (iii) The chess reference-rig. (iv) The active-camera marker system. The exploitation of these elements partially provides the 6D-pose ground truth of the camera frame and the 6D-pose of the reference-rig(s). The aim of use different patterns is to analyze the differences in terms of uncertainty behaviors

Head markers: On the humanoid robot head, three spherical markers were placed in a non-collinear arrangement. They are used to estimate the head kinematic frame $H \in SE^3$.

Reference-rigs: The blob-rig is made of two overlapped patterns: (i) The 2D pattern consisting of three black asymmetrically distributed circles with rather large diameter ø 60 mm. (ii) The 3D pattern consisting of three spherical markers respectively placed at the centers of the circles. The alignment between both patterns is done by a translation along the 3D normal $N_{\text{Blobs}} \in \mathbb{R}^3$ of the blobs pattern, namely the offset to the marker center. Hence, the reference-rig has two kinematic frames, the χ-printed kinematic frame and the ς-spherical markers frame. Second, the chess-rig is a 800×600mm standard reference pattern with three reference markers located at the corners. This chess reference-rig also involves the two kinematic frames χ and ς with an alignment vector $N_{\text{Chess}} \in \mathbb{R}^3$, namely the translation vector from the chess-pattern center point to the spherical markers frame. Pattern dimensions are determined for the visual recognition at wide-depths further than 3,500 mm using a 6 mm lenses with VGA images.

Labeled marker positions: The multiple view fusion with active-cameras is done by the marker system. All marker positions are calculated relatively to the world coordinate system $W \in SE^3$ established at the initial calibration (see Fig. 6.4).

Autonomous uncertainty sampling: In order to obtain highly representative observation samples, the distributed locations and orientations of the humanoid head for visual recognition are planned and controlled by a scanning plan (see two scanning plans in Figs. 6.4 and 6.5). This is done by transversing the path of sampling nodes.

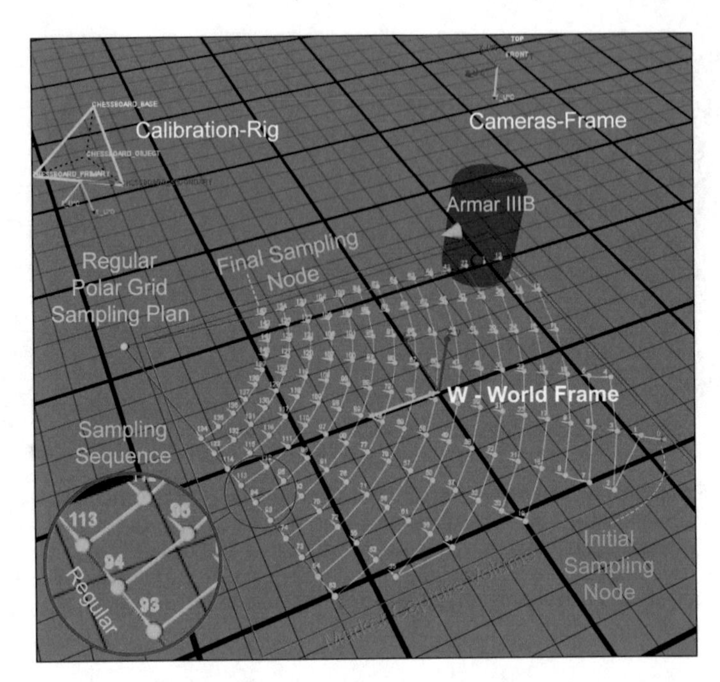

Fig. 6.4 Autonomous uncertainty sampling. The red rectangle on the floor is the boundary of the marker system. The sampling plan is a distributed set of linked 3D poses (2D for location and 1D for orientation of the robot platform) also called *sampling nodes*. A sampling node includes a set of neck configurations. This regular polar grid plan is direct and intuitive but naive and faulty. This produces sampling artifacts because the nodes along each arc are located at the same distance from the reference-rig. This plan generates isolated sample clusters preventing the proper depth generalization by kernel density estimation

Within each node, various configurations of the humanoid robot neck (pitch, roll and yaw) are reached. This introduces the actuators uncertainty effects in the samples. Furthermore, at each neck configuration, several recognition trials were performed to obtain highly representativeness of the configuration. The coordinated acquisition ensures the sound data association between the visual perception of the robot and the marker reference system in terms of temporal consistency, spatial uniformity and active sensing representativeness by the inclusion of uncertainty effects of the head actuators. The systematic generation and execution of scanning plans has various advantages. First, because the process is totally autonomous, it was possible to extensively collect large amount of learning samples (in the experiments more than 25,000 registered observations were collected) allowing the generation of high quality uncertainty models. In addition, the systematic scanning plans enable the comparison of extraction methods.

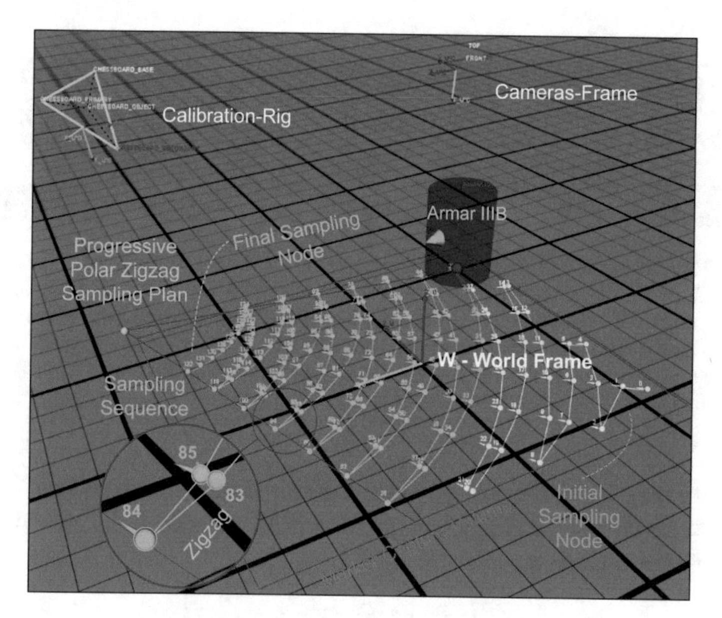

Fig. 6.5 The progressive polar zigzag plan enables the proper acquisition of uncertainty observations. The sampling nodes are located in a regular depth progression with adjusted angular distribution (see resulting model in Fig. 6.8)

6.5 Visual Measurements

Blob reference-rig: The humanoid robot visually estimates the 6D-pose of the blob reference-rig as follows. In order to segment the pattern circles, the input color image $I_{RGB}(\mathbf{x}) \in \mathbb{N}^3$ is transformed to a normalized saliency image by $I_S(\mathbf{x}, \tau) = 1 - \frac{1}{2^{3m\tau}} [I_R(\mathbf{x}) \cdot I_G(\mathbf{x}) \cdot I_B(\mathbf{x})]^{\tau} \in \mathbb{R}$, where $\tau \in \mathbb{N}^+$ improves the contrast between circles and background and m is the bits per pixel (Eq. 4.2) (see Fig. 6.6). After, the mask image $I_A(\mathbf{x}) \in \{0, 1\}$ was computed by adaptive thresholding [13] (Fig. 6.6b). Next, a region growing algorithm extracts the blobs B_i. The centroid of each blob $\bar{\mathbf{x}}_i$ is estimated by integrating its radial weighted saliency. The blobs in the left B_i^L and right B_k^R image were matched using epipolar geometry [8]. Matches with low confidence or outside the depth interval ($\delta_0 = 500 \leq |X_i| \leq \delta_1$=3,500 mm) were removed. Based on the 3D positions of the blobs X_i^v (the superindex v denotes vision estimated), the marker correspondence was performed using the center distances. Finally, the matched blob centers X_B^v, X_E^v and $X_S^v \in \mathbb{R}^3$ described the kinematic transformation $T_C^{\chi^v}$ from the camera kinematic frame C to the χ-printed pattern as an homogeneous matrix

$$M^v = X_E^v - X_B^v,\ N^v = M^v \times (X_S^v - X_B^v),\ P^v = M^v \times N^v$$
$$T_C^{\chi^v} = \begin{bmatrix} \widehat{M}^v & \widehat{N}^v & \widehat{P}^v & X_B^v \\ 0 & 0 & 0 & 1 \end{bmatrix}, \tag{6.1}$$

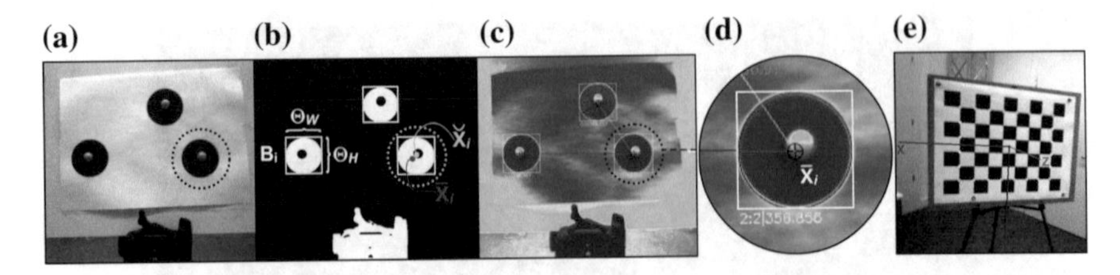

Fig. 6.6 The humanoid robot visual system recognizes and estimates the 6D-pose of the reference-rigs. In **a–d**, results of the processing pipeline at a close distance (∼356.8 mm). **a** The input color image $I_{RGB}(\mathbf{x})$. **b** The mask image $I_A(\mathbf{x})$. **c** The saliency image $I_S(\mathbf{x}, \tau)$ with labeling and the estimated depth. **d** The zoom circle shows the recognized blob with its matching identifier and the visual depth measurements. **e** The kinematic frame of the chess reference-rig is robustly done by Eigen-decomposition of corner points

where $\widehat{\cdot}$ denotes the unitary vector. Finally, the kinematic transformation $\mathcal{T}_C^{\varsigma^v}$ from the camera kinematic frame C to the spherical markers frame ς was attained by the alignment offset N_{Blobs} as

$$\mathcal{T}_C^{\varsigma^v} = \begin{bmatrix} \widehat{M}^v & \widehat{N}^v & \widehat{P}^v & (X_B^v + N_{\text{Blobs}}) \\ 0 & 0 & 0 & 1 \end{bmatrix}. \tag{6.2}$$

This is the transformation from the camera kinematic frame C to the reference-rig. This is the key to bidirectionally relate the ground truth system with the visual depth perception.

Chess reference-rig: The extraction of corner points from the chess reference pattern is a well studied problem. It has been properly solved (see [14]). The extracted noisy 3D corner points P_i^v were used to determine their centroid $\overline{P^v}$ corresponding to the base of the χ kinematic frame of the chess reference-rig. Finally, from the covariance matrix $G^v = \frac{1}{n}\sum_{i=1}^{n}(P_i^v - \overline{P^v})(P_i^v - \overline{P^v})^T$ using SVD decomposition,[2] the Eigenvector with smallest associated Eigenvalue is the normal of the plane. The other axes are determine using the geometry of the pattern. The representation of this kinematic frame is analogous to Eq. 6.2 (see Fig. 6.6e).

6.6 Ground Truth Measurements

In order to sample the uncertainty distribution of the visual depth, all kinematic frames have to be linked into a kinematic tree (see Fig. 6.3). This enables global unified temporal association of the 6 DoF measurements obtained based on both systems. The linkage arises from the following bidirectional transformations:

[2]Since G^v is a 3×3 matrix, the the singular values and vector coincide with the Eigenvalues and Eigenvectors.

- **World to head**: The labels and position of the markers are obtained by a network interface. These measurements have submillimeter accuracy according to [7]. The kinematic frame on top of the humanoid robot head $\mathcal{T}_W^{H^m}$ is calculated by Eq. 6.1 where the superindex m denotes the source of information is the marker system.
- **World to reference-rig**: The transformation $\mathcal{T}_W^{\varsigma^m}$ is also computed (as in Eq. 6.1) using the marker positions and labels.
- **Reference-rig to cameras**: The transformation from the kinematic frame on the reference-rig to the camera kinematic frame is the inverse of Eq. 6.2, namely $\mathcal{T}_{\varsigma^v}^C = [\mathcal{T}_C^{\varsigma^v}]^{-1}$.
- **World to camera**: This transformation results from the forward kinematic chain of previous two transformations as $\mathcal{T}_W^C = \mathcal{T}_{\varsigma^v}^C \mathcal{T}_W^{\varsigma^m}$. This is the coupling of the visual perception ς^v to the marker ground truth ς^m, namely the connection from the χ-printed kinematic frame to the spherical markers frame ς.
- **Head to camera**: Results from the kinematic chain coupling the inverse transformation from world to head and the direct transformation from the world to camera is expressed as $\mathcal{T}_{H^m}^C = \mathcal{T}_W^C [\mathcal{T}_W^{H^m}]^{-1}$. This is the camera registration relative to the head kinematic frame. In order to accurately achieve this registration, the humanoid robot should be close to the reference-rig while performing this procedure (see Fig. 6.6 a–d). Once attained, this transformation is fixed and stored for the sampling process. Thus, when the humanoid robot moves, the transformation from the world kinematic frame W to the camera kinematic frame C is properly determined as $\mathcal{T}_W^C(t) = \mathcal{T}_{H^m}^C \mathcal{T}_W^{H^m}(t)$, where the t argument is the time stamp. This dynamic transformation unifies the visual perception and the ground truth measurements (Fig. 6.7).

6.7 Uncertainty Distribution Learning

A learning observation of the visual depth uncertainty is composed as

$$S_t := \Big[\underbrace{\delta := \Phi\Big(\mathcal{T}_C^{\varsigma^v}(t)\Big)}_{\delta\text{-visual depth (Eq. 6.2)}} \;,\; \underbrace{\gamma := \Phi\Big(\mathcal{T}_C^{\varsigma^m}(t)\Big)}_{\gamma\text{-ground truth depth (Eq. 6.2)}} \Big]^T \in \mathbb{R}^2, \tag{6.3}$$

where the depth function $\Phi : SE^3 \mapsto \mathbb{R}$ extracts the displacement length of the transformation. The observation S_t (see Fig. 6.8) integrates the vision $\mathcal{T}_C^{\varsigma^v}(t)$ from Eq. 6.2 and marker information

$$\mathcal{T}_C^{\varsigma^m}(t) = \mathcal{T}_W^{\varsigma^m}(t)[\mathcal{T}_W^C(t)]^{-1}. \tag{6.4}$$

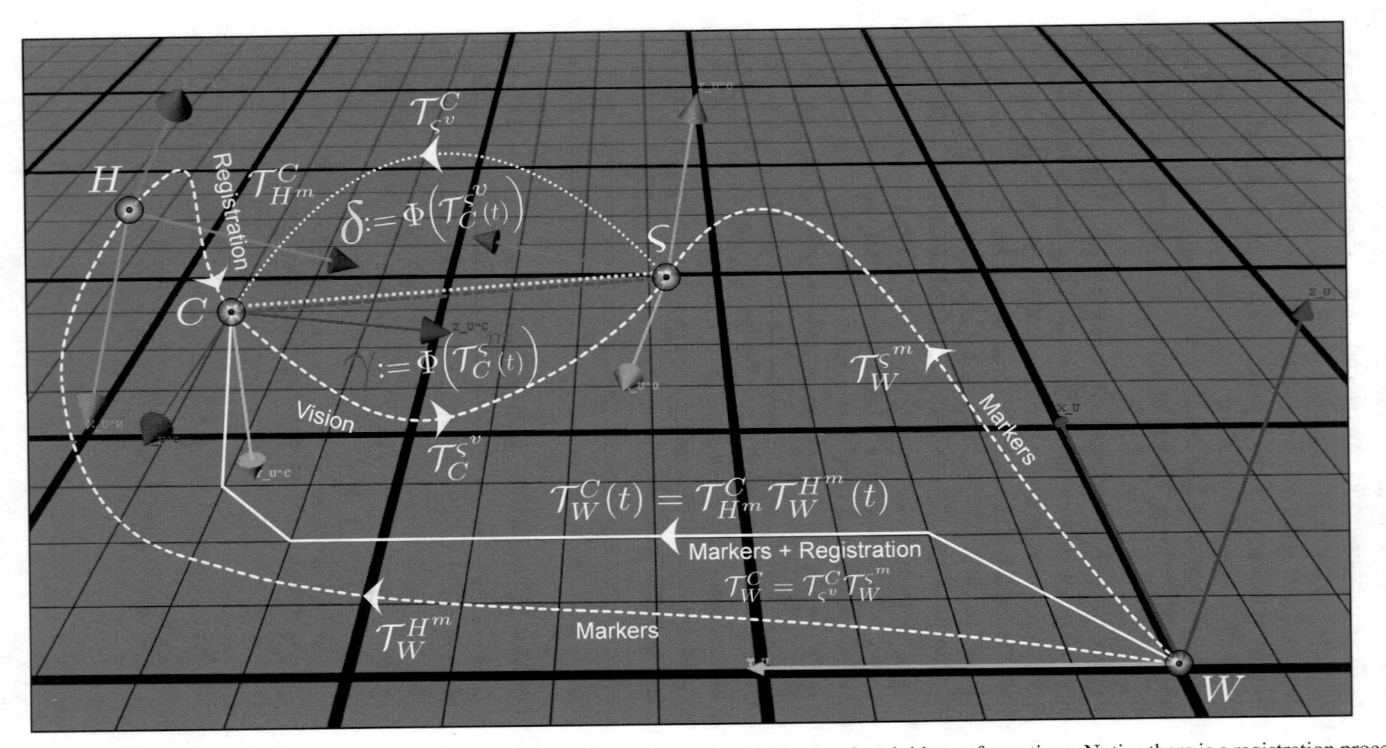

Fig. 6.7 The ground truth kinematic tree connects all the kinematic frames by means of time varying rigid transformations. Notice there is a registration process in order to attain the fixed transformation from the head kinematic frame H to the cameras kinematic frame C. The perceptual depth δ and the ground truth depth γ are shown emphasizing the source frames ς^v and ς^m

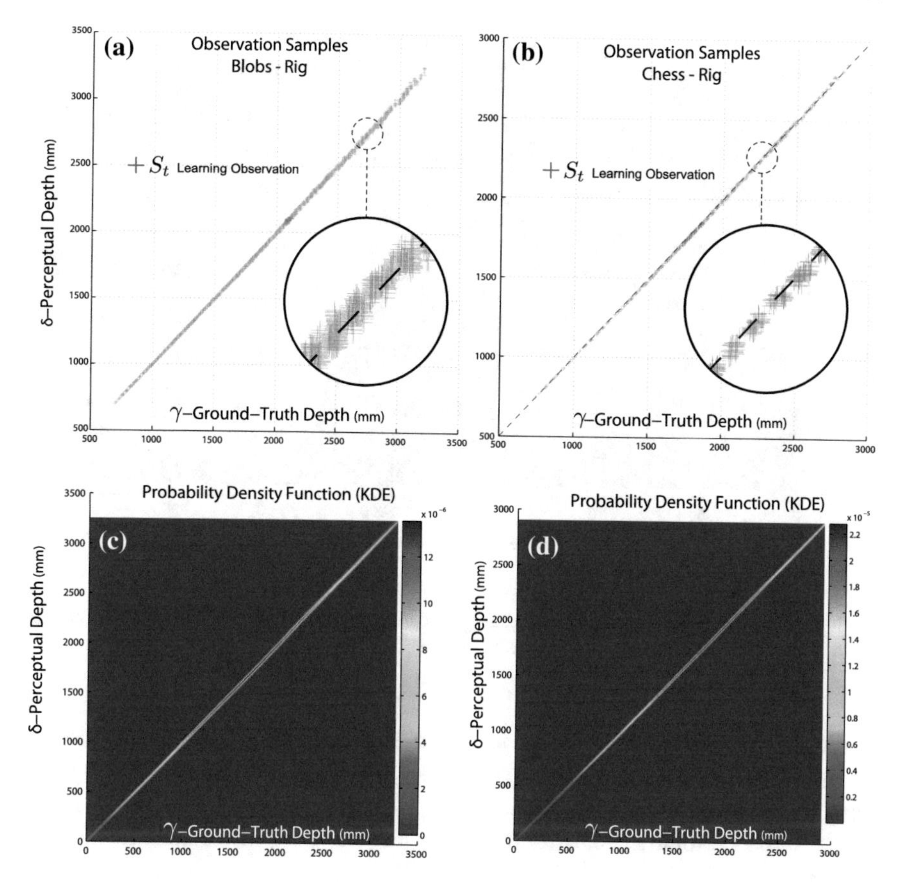

Fig. 6.8 **a** and **b** The show 2D learning observations S_t (Eq. 6.3) relating visual depth perception δ to ground truth depth γ. **c** and **d** The shows the learned models by KDE using the adaptive band-width (Eq. 6.5)

6.7.1 Model Learning

The uncertainty PDF of the depth perception Ψ is a random variable function sampled by a collection of learning observations using Eq. 6.3. The inference based on the sampling set $L := \{S_t\}_{t=1}^m$ is done by kernel density estimation [15]. The continuous model implies that for a perceptual depth δ there is a corresponding uncertainty distribution ζ such as $\forall\ (\delta_0 \leq \delta \leq \delta_1)\ \exists\ \Psi(\delta) \Rightarrow\ \zeta(\delta, 0 \leq \gamma < \infty)$ and consequently $\int_0^\infty \zeta(\delta, \gamma)\, d\gamma = 1$.

The distribution Ψ is inferred as (see Fig. 6.8)

$$\zeta(\delta, \gamma) = \sum_{t=1}^{m} K_{\lambda(\delta,\gamma)}\Big(||[\delta, \gamma]^T - S_t|| \Big), \tag{6.5}$$

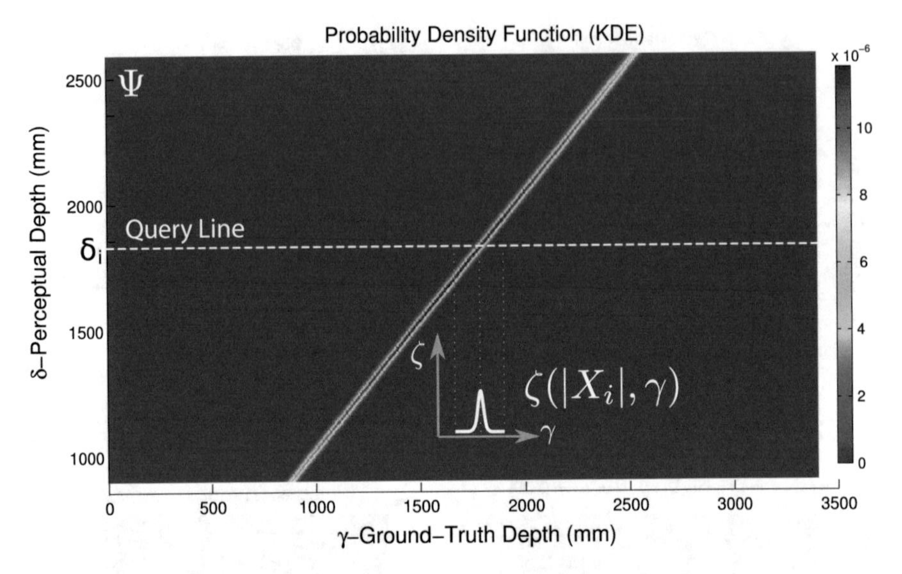

Fig. 6.9 The concept of depth query. The uncertainty model Ψ enables efficient computation of depth queries $\varsigma(|X_i|, \gamma)$. This mechanism equips the humanoid robot with means to determine the depth probability distribution of a visually estimated point in space

where the Gaussian kernel $K_{\lambda(\delta,\gamma)}$ and its adaptive bandwidth are determined by the generalized Scott rule [16] as $\lambda(\delta, \gamma) = hm^{-\frac{1}{6}}\widehat{\Sigma}^{\frac{1}{2}}(\delta, \gamma)$, where $\widehat{\Sigma}(\delta, \gamma)$ is the sampling covariance matrix centered at $[\delta, \gamma]^T$, m denotes the cardinality of L and $h \in \mathbb{N}^+$ is an isotropic scale factor for the bandwidth matrix (see detailed evaluation in [17]).

The learned function $\varsigma(\delta, \gamma)$ is a non-parametric continuous distribution sampled with fine discretization (κ=1 mm) into a lookup table $\tilde{\varsigma}(\delta, \gamma)$ with $\lceil (\frac{\delta_1 - \delta_0}{\kappa}) \rceil^2 = (3460)^2 = 11,971,600$ support points (see Fig. 6.8).

6.7.2 *Model Analysis*

The analysis of the uncertainty model Ψ is based on the concept of *depth query* $\varsigma(|X_i|, \gamma)$. When a point in space X_i has been estimated by the visual perception of the humanoid robot, it is possible to obtain its depth uncertainty distribution $\Psi(|X_i|) \mapsto \varsigma(|X_i|, \gamma)$, which efficiently, compactly and non-parametrically describes the visual depth PDF along the ground truth depth γ. This inference depth query $\varsigma(|X_i|, \gamma)$ results from the learned ground truth model (without assumptions or reductions) of the visual perception process (see Fig. 6.9).

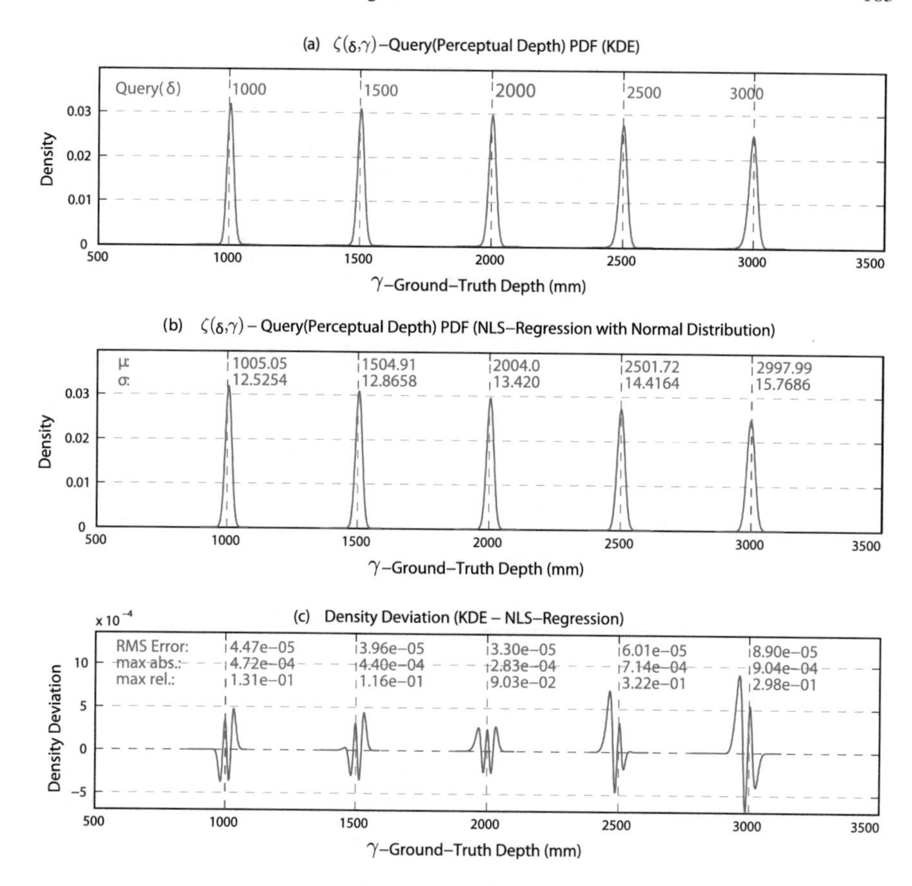

Fig. 6.10 Queries of the ground truth uncertainty model of depth perception Ψ using the blob-pattern. **a** Five different examples of queries show the inferred functions $\zeta(\delta, \gamma)$ (Eq. 6.5). These curves are learned uncertainty profiles. **b** Resulting curves by NLS (with normal distribution $\mathcal{N}(\mu_\delta, \sigma_\delta) \cong \Lambda_{\text{NLS}}(\zeta(\delta, \gamma))$ from the upper plot. **c** The RMS-deviation between non-parametric model and normal regression curves (with maximal RMS deviation of 0.0089% @ 3000 mm depth) support the use of normal distribution as function of the depth

Parametric Uncertainty Model: Normal Distribution Plausibility

A study of the uncertainty model Ψ was conducted on regular distributed queries applying nonlinear least squares (NLS) regression $\mathcal{N}(\mu_\delta, \sigma_\delta) \cong \Lambda_{\text{NLS}}(\zeta(\delta, \gamma))$ [18] with depth intervals of κ=1 mm (see Figs. 6.10 and 6.11).

In the NLS-regression, the normal distribution profile was used as the underlying regression shape in order to compare the parametric distribution with the non-parametric approach. This obtains the systematic-error curve (mean *vs* depth), the stochastic-spreading curve (standard deviation *vs* depth) and their RMS-deviation curve (normal-plausibility *vs* depth). The resulting regression curves were compared against the non-parametric queries (see results in Figs. 6.12 and 6.13). These find-

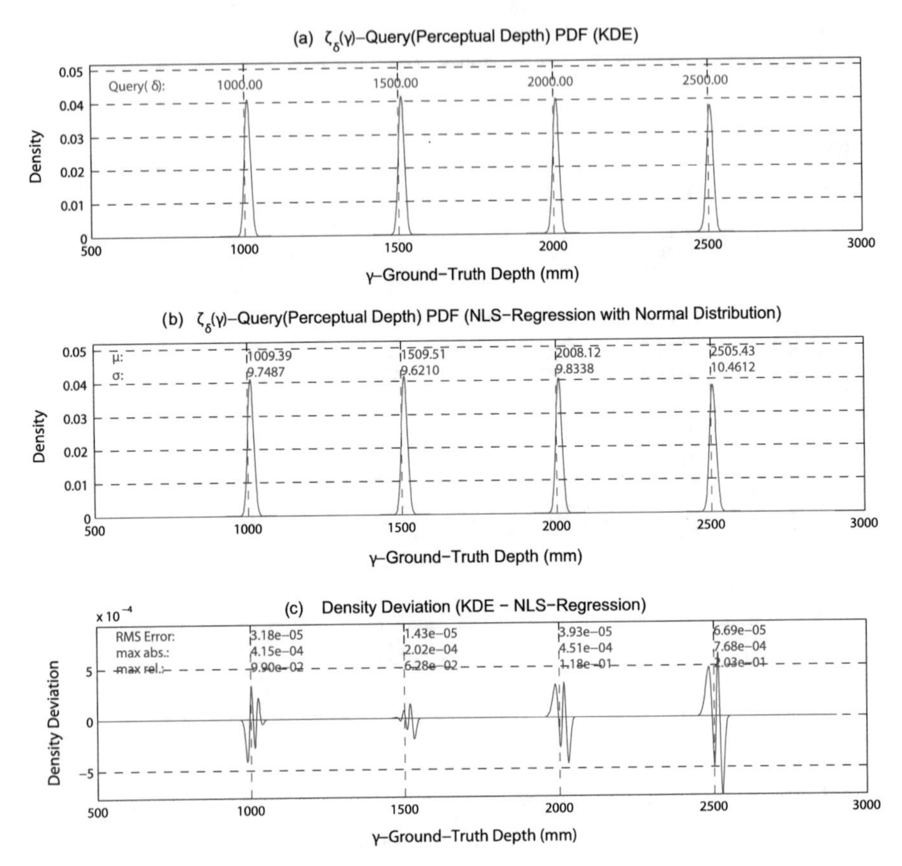

Fig. 6.11 Queries of the ground truth uncertainty model of depth perception Ψ using chess-pattern. **a** Queries show the inferred functions $\zeta(\delta, \gamma)$ (Eq. 6.5). **b** Resulting curves by NLS. **c** The RMS-deviation between the non-parametric model and the normal regression curves (with a maximal RMS deviation of 0.0069% @ 3000 mm depth) also support the use of normal distribution as a profile of the depth uncertainty

ings yield to the extraction of an optimal (compact and efficient) representation in the following section.

Efficient Visual Uncertainty Depth Model: High Order Polynomials

The normal distribution from NLS-regression at the depth $\delta_i = \delta_0 + i\kappa$ enables its use only at this particular depth. In order to generalize this concept for the whole learned depth, a polynomial fitting of the normal distribution parameters is conducted, namely the mean $\mu_\Psi(\delta)$ and standard deviation curves $\sigma_\Psi(\delta)$ through polynomial regression expressed as

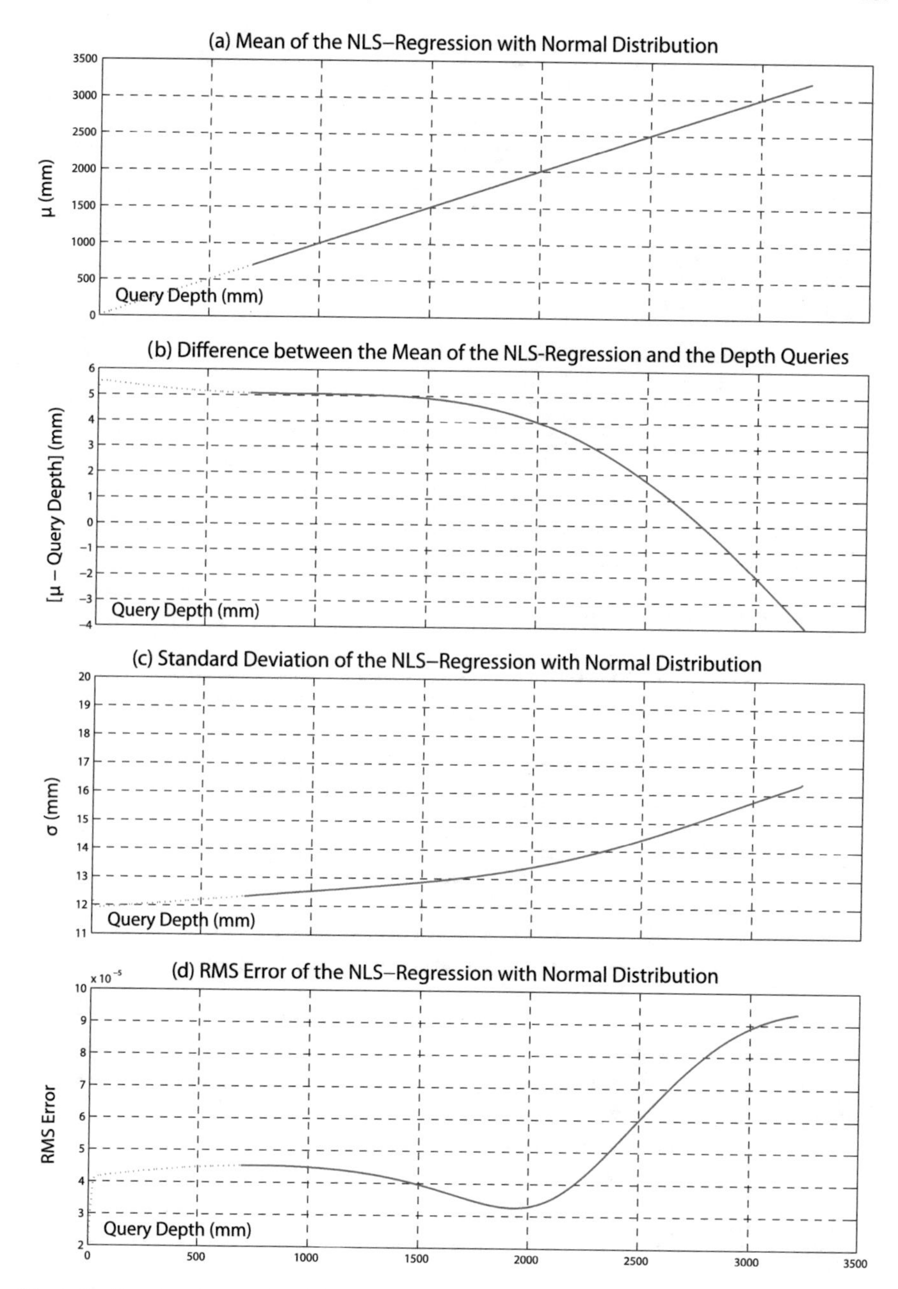

Fig. 6.12 Comparative results of the normal distribution plausibility using NLS-regression on the non-parametric model Ψ from the blobs reference-rig

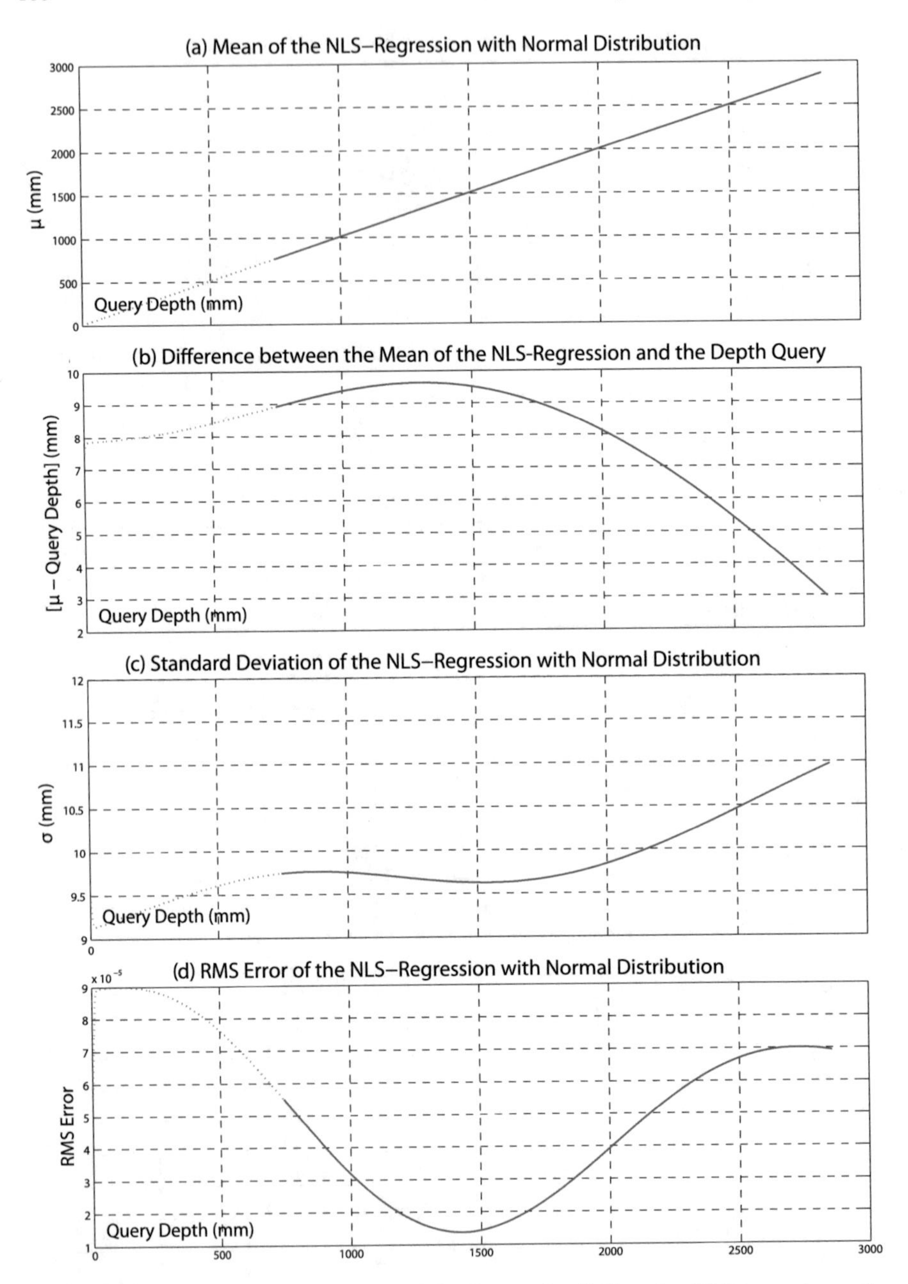

Fig. 6.13 Comparative results of the normal distribution plausibility using NLS-regression on the non-parametric model Ψ from the chess reference-rig

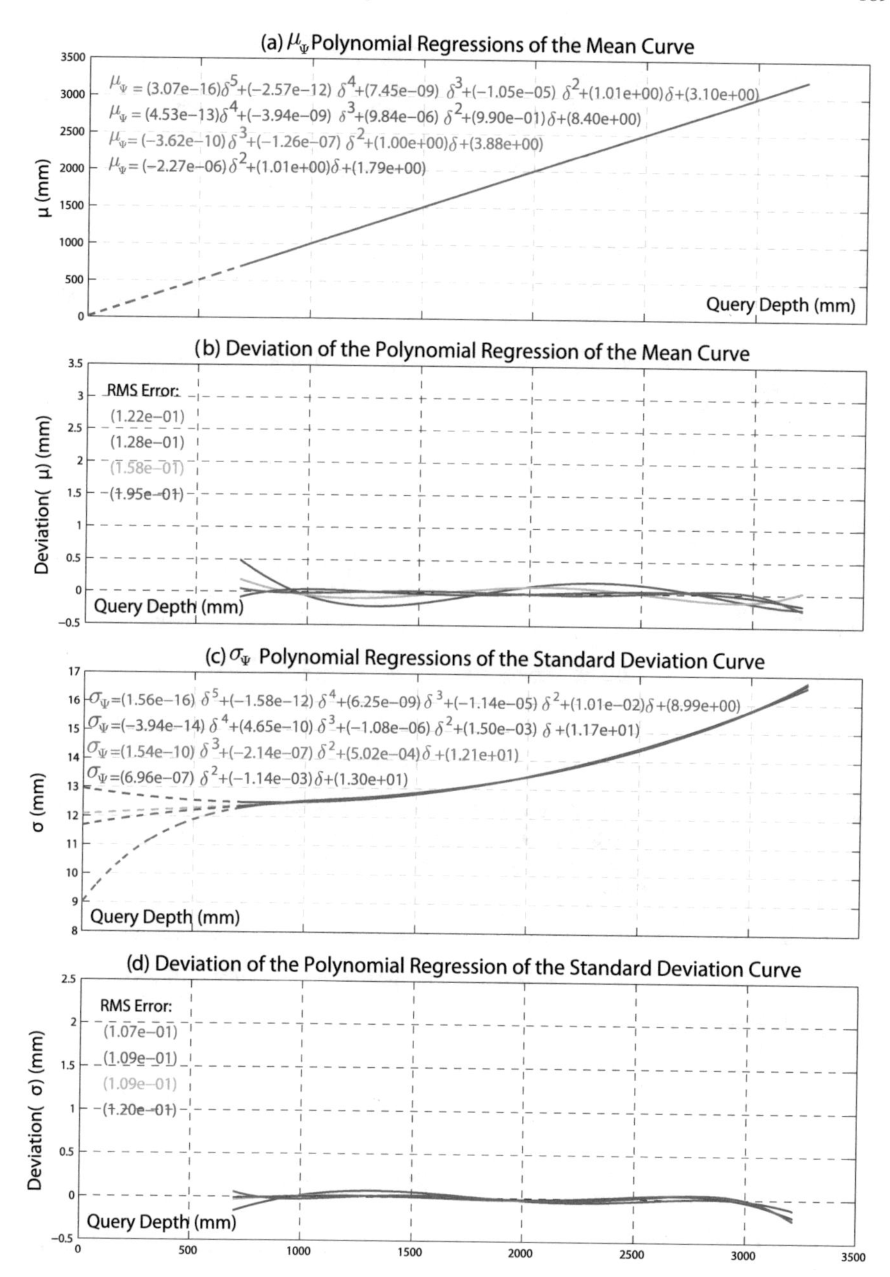

Fig. 6.14 Polynomial representation of the normal distribution parameters from the NLS-regression of Ψ (in Sect. 6.7.2) using the blobs reference-rig. These expressions provide the mean $\mu_\Psi(\delta)$ and standard deviation $\sigma_\Psi(\delta)$ as functions of the continuous depth δ in consistent and efficient manner

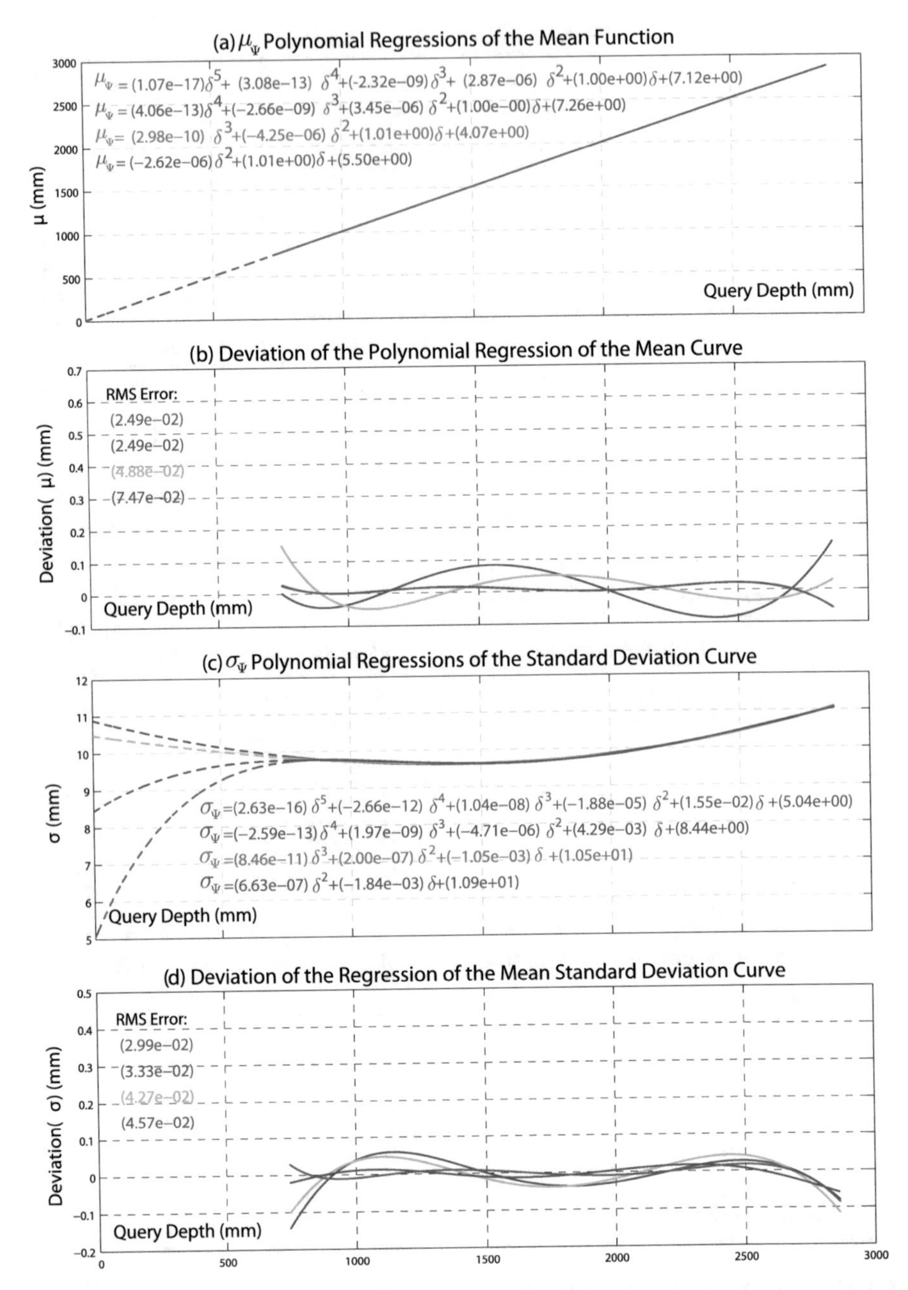

Fig. 6.15 Polynomial representation of the normal distribution parameters from the NLS-regression of Ψ (in Sect. 6.7.2) using the chess reference-rig. These polynomial reflect the mean $\mu_\Psi(\delta)$ and standard deviation $\sigma_\Psi(\delta)$ depending on the continuous depth δ by a computational efficient representation

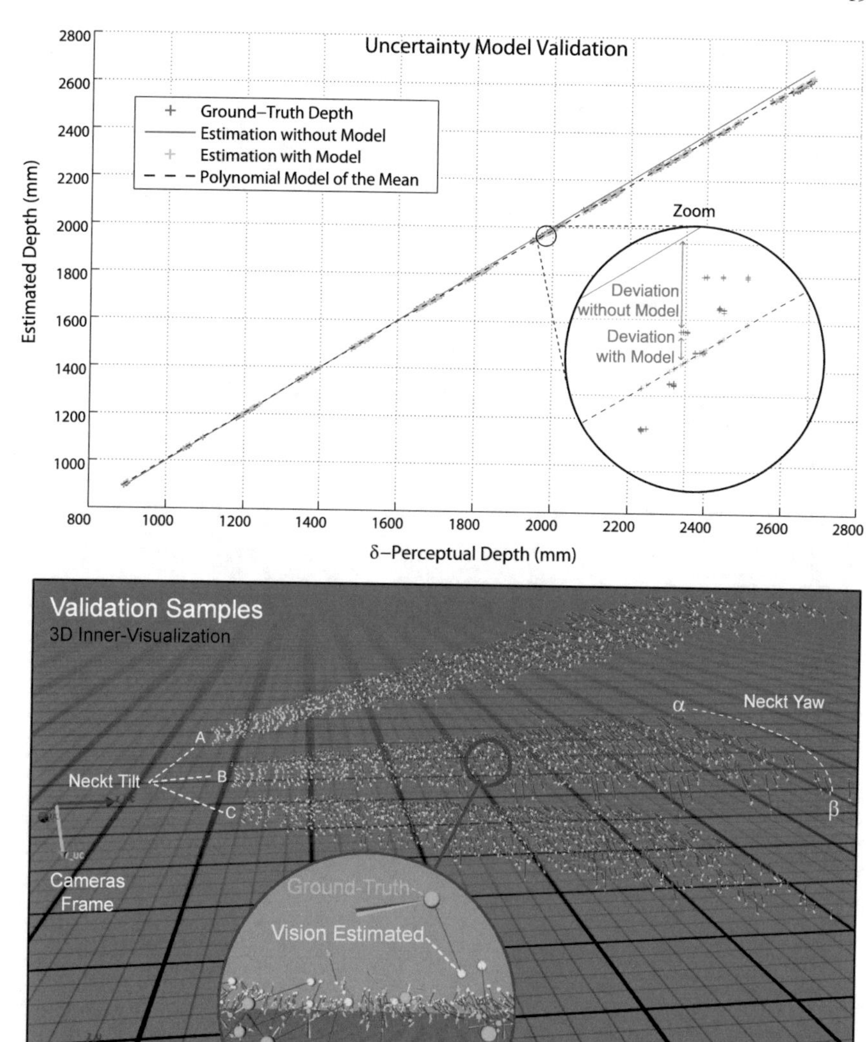

Fig. 6.16 The validation of the model was realized by comparing the visual and predicted depths from the uncertainty model and the real depth attained by the marker system (see the relative and absolute errors compared with the estimated standard deviation in Fig. 6.17)

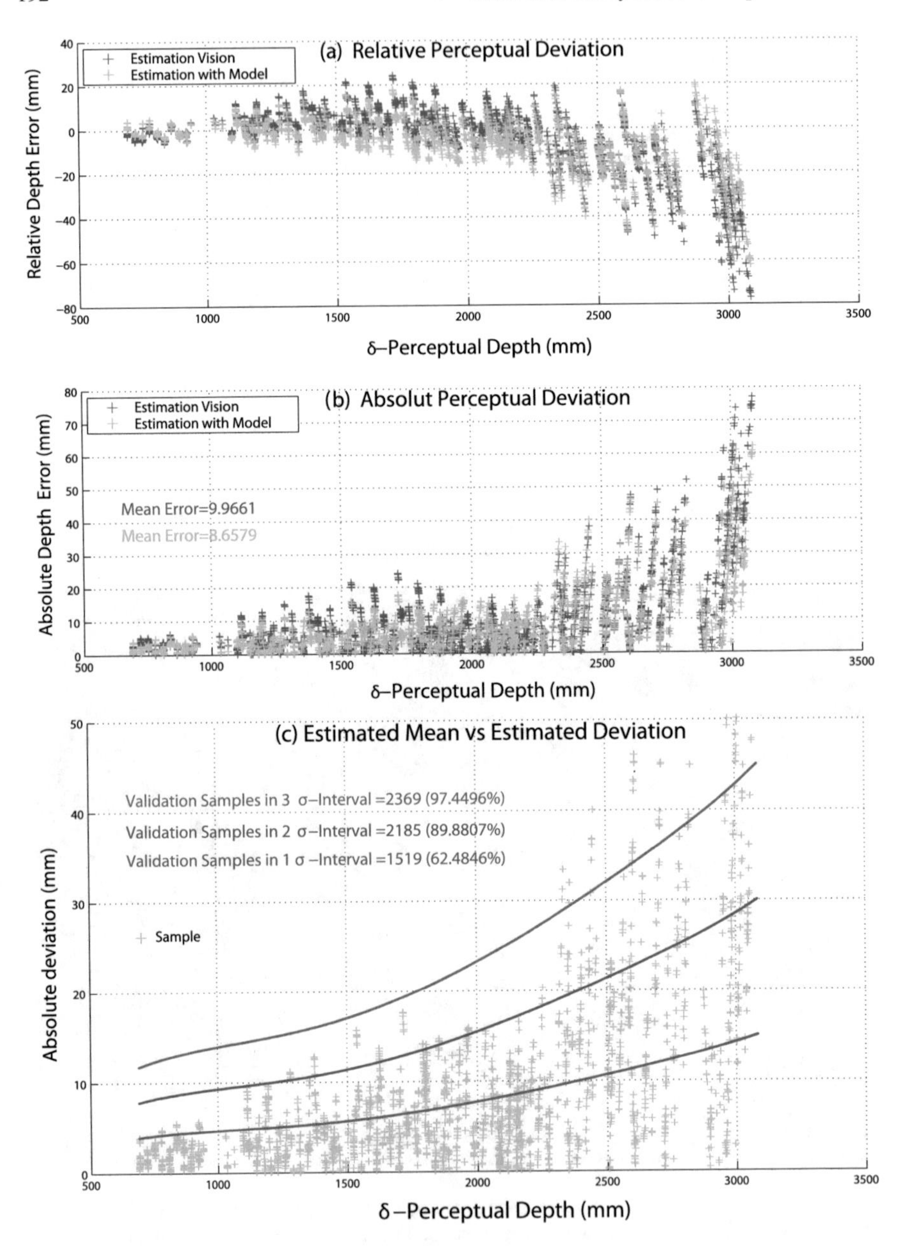

Fig. 6.17 The absolute and relative depth deviations analyzed according the mean and standard deviation curves from the uncertainty model (see Figs. 6.14 and 6.15)

$$\underbrace{\begin{bmatrix} \delta_{\kappa_1}^n & \delta_{\kappa_1}^{n-1} & \ldots & \delta_{\kappa_1} & 1 \\ \vdots & \vdots & \ddots & \vdots & \vdots \\ \delta_{\kappa_m}^n & \delta_{\kappa_m}^{n-1} & \ldots & \delta_{\kappa_m} & 1 \end{bmatrix}}_{X_n} \underbrace{\left[a_n, a_{n-1}, \ldots, a_1, a_0\right]^T}_{A_n(\mu_\Psi)} = \underbrace{\left[\mu_1, \ldots, \mu_m\right]^T}_{Y_m(\mu_\Psi)},$$

$$A_n(\mu_\Psi) = X_n^* Y_m(\mu_\Psi) \;\Rightarrow\; \mu_\Psi(\delta) = \sum_{i=0}^{n} a_i \delta^i, \tag{6.6}$$

where the X_n^* denotes the pseudo inverse and $m = \lceil (\frac{\delta_1 - \delta_0}{\kappa}) \rceil = 3460$ (see the resulting polynomial mean curve $\mu_\Psi(\delta)$ in Fig. 6.14). For the standard deviation the same formulation $A_n(\sigma_\Psi) = X_n^* Y_m(\sigma_\Psi) \Rightarrow \sigma_\Psi(\delta) = \sum_{i=0}^{n} a_i \delta^i$ can be applied (see the result in Fig. 6.15). These analyses include various ($2 \leq n \leq 5$) orders[3] of the polynomial expressions and their corresponding deviations compared to the Ψ model using NLS-regression of depth queries (as in Sect. 6.7.2).

6.8 Experimental Model Validation

The validation of the proposed uncertainty model was performed by collecting new samples S_t (Eq. 6.3) and comparing the predictions provided by the learned uncertainty model in terms of polynomial functions of the depth $\mu_\Psi(\delta)$ (Fig. 6.14) and $\sigma_\Psi(\delta)$ (Fig. 6.15). The validation (see Fig. 6.16) evidences that the proposed model Ψ can be used to soundly reflect the perceptual uncertainty of the visual landmarks by estimating the correction of the depth perception (compensation of the systematic-error) while simultaneously determining the accurate stochastic-spreading by high oder polynomial functions (see the summary validation in Fig. 6.17).

6.9 Discussion

The contribution of this chapter is a novel ground truth based uncertainty model of the depth perception for humanoid robots. The proposed method conveniently overcomes the analytical limitations of previous models while integrating all uncertainty sources. The setup and the developed recognition methods exploit the accurate ground truth attained by the marker system. The systematic collection of perceptual samples allows to obtain high quality models in a fully automatic fashion. The attained uncertainty models (presented in Figs. 6.10 and 6.11) corroborate the motivation of the approach. These results support the need to learn the uncertainty model of the

[3] Higher orders $n \geq 6$ were analyzed resulting insignificant deviations.

applied recognition mechanism. Furthermore, the representation and exploitation of the model into a lookup table and polynomial functions are appropriate techniques for wide range of real-time visual perception applications. A unique and remarkable property of the proposed uncertainty model is its capacity to correct the systematic depth errors in the visual depth perception. This novel property together with the coherent estimation of the standard deviation at each arbitrary and continuous depth are the successfully achieved objectives of this chapter. However, the learned models expose high dependency on the stereo calibrations. This has been extensively analyzed in [17]. This is a critical point of the method. However, the fact that visual depth uncertainty of the humanoid robot can be properly represented (with polynomial expression of degree five) enables the fast learning by considering only a few samples (see Sect. 8.2). The focus in this chapter is placed on the depth deviation directly formulated by the depth function Φ (in Eq. 6.3). This function maps the learning space from 6D-poses to the 1D visual depth subspace. Because the uncertainty of the 6D-pose was extensively and properly sampled, it can be exploited more widely using the observations S_t (Eq. 6.3) and a more general 6D-mapping Φ' (see results in [17]). Additionally, the time varying kinematic tree formulation of the humanoid robot and the elements in the environment could also be used in many other contexts and applications (SLAM, robot-machine interaction, motion graphs, imitation, validation of vision algorithms and kinematic learning methods) for robots were the marker positions and robot configurations are unified in a spatio-temporal reference frame.

References

1. Alvertos, N. 1988. Resolution Limitations and Error Analysis for Stereo Camera Models. In *IEEE Conference Southeastcon*, 220–224.
2. Miura, J., and Y. Shirai. 1993. An Uncertainty Model of Stereo Vision and its Application to Vision-Motion Planning of Robot. In *International Joint Conference on Artificial Intelligence*, 1618–1623.
3. Sinz, H., J. QuiÃonero, G. Bakir, C. Rasmussen, and M. Franz. 2004. Learning Depth from Stereo. In *on PatternRecognition*, ed. D.A.G.M. Symposium, 245–252. Berlin: Springer.
4. Swapna, P., N. Krouglicof, and R. Gosine. 2009. The Question of Accuracy with Geometric Camera Calibration. In *Canadian Conference on Electrical and Computer Engineering*, 541–546.
5. Perrollaz, M., A. Spalanzani, and D. Aubert. 2010. Probabilistic Representation of the Uncertainty of Stereo-vision and Application to Obstacle Detection. In *IEEE Intelligent Vehicles Symposium*, 313–318.
6. Di Leo, G., C. Liguori, and A. Paolillo. 2011. Covariance Propagation for the Uncertainty Estimation in Stereo Vision. *IEEE Transactions on Instrumentation and Measurement* 60 (5): 1664–1673.
7. Asfour, T., K. Regenstein, P. Azad, J. Schröder, A. Bierbaum, N. Vahrenkamp, and R. Dillmann. 2006. ARMAR-III: An Integrated Humanoid Platform for Sensory-Motor Control. In *IEEE RAS International Conference on Humanoid Robots*, 169–175.
8. Hartley, R., and A. Zisserman. 2004. *Multiple View Geometry in Computer Vision*, 2nd ed. Cambridge: Cambridge University Press. ISBN 0521540518.

9. Pollefeys, M., L. Van Gool, M. Vergauwen, F. Verbiest, K. Cornelis, J. Tops, and R. Koch. 2004. Visual Modeling with a Hand-Held Camera. *International Journal of Computer Vision* 59 (3): 207–232.
10. Noskovicova, L., and R. Ravas. 2010. *Subpixel Corner Detection for Camera Calibration*, 78–80. International Symposium: In MECHATRONIKA.
11. Okada, K., M. Kojima, S. Tokutsu, T. Maki, Y. Mori, and M. Inaba. 2007. Multi-cue 3D Object Recognition in Knowledge-based Vision-guided Humanoid Robot System. In *IEEE-RSJ International Conference on Intelligent Robots and Systems*, 3217–3222.
12. Azad, P., T. Asfour, and R. Dillmann. 2009. Combining Harris Interest Points and the SIFT Descriptor for Fast Scale-Invariant Object Recognition. In *IEEE-RSJ International Conference on Intelligent Robots and Systems*, 4275–4280.
13. Bradley, D., and G. Roth. 2007. Adaptive Thresholding using the Integral Image. *Journal of Graphics, GPU, and Game Tools* 12 (2): 13–21.
14. KIT. 2011. Karlsruhe Institute of Technology, Computer Science Faculty, Institute for Anthropomatics. The Integrating Vision Toolkit. http://ivt.sourceforge.net.
15. Duda, R., P. Hart, and D. Stork. 2001. *Pattern Classification*, 2nd ed. New York: Wiley. ISBN 978-0471056690.
16. Scott, D. 1992. *Multivariate Density Estimation: Theory, Practice, and Visualization*. Wiley Series in Probability and Mathematical Statistics: Applied Probability and Statistics. New York: Wiley.
17. Vollert, M. 2012. *Visual ground truth: Modellierung der wahrnehmungsunsicherheit in einem stereokamerasystem*. KIT, Karlsruhe Institute of Technology, Computer Science Faculty, Institute for Anthropomatics: Technical Report.
18. Bates, D.M., and D.G. Watts. 1988. *Nonlinear Regression and Its Applications*. New York: Wiley. ISBN 978-0471816430.

Chapter 7
Global Visual Localization

A formal representation of the elements composing the surroundings and their spatial interrelations is needed to enable robots to perform complex tasks through the composition of multimodal skills accomplished in a perception-action cycle (see Chaps. 1 and 2) (Fig. 7.1).

An effective mechanism to achieve the self-localization in these environments must profit from the intrinsic topological and geometric structure of the world by exploiting invariant properties of the environmental elements (see Chap. 5). This mechanism has to sagaciously face many diminishing factors that complicate the self-localizing task, namely the granularity of the model, the nature of the sensors (Chap. 4) and the uncertainty of the perception-recognition cycle (Chap. 6). Ideally, the visually extracted geometric features (Chap. 5) contain necessary information to obtain the vision to model coupling transformation, namely the 6D-pose of the robot. Notice that the estimation of the pose is affected by uncertainties introduced during the recovering of euclidean metric from images as well as at the mapping from the camera to the ego-frame.[1] The proposed method models the uncertainty of the percepts with a radial density distribution resulting from the plausibility analysis in Sect. 6.7.2. The formulation based on density spheres allows a closed-form solution which not only derives the maximal density position depicting the optimal ego-frame but also ensures a solution even in situations where pure geometric spheres might not intersect (see Sect. 7.1). The natural usage of concepts from geometric algebra [1] arises from the key idea of applying restriction subspaces in order to constraint and find the location of the robot. In this manner, the formulation profits from mathematical frameworks (geometric algebra and computational geometry), for instance, the generalized intersection operator $\wedge$ of geometric entities such as planes, lines, spheres, circles, point pairs and points. This so-called wedge operator is the ideal

---

[1] The reference frame of the humanoid robot which remains static during the execution of the active visual scan for global localization. Depending on whether the scan strategy includes the hip, this ego-frame is usually located at the platform- or shoulders-frame of the humanoid robot.

D. I. González Aguirre, *Visual Perception for Humanoid Robots*,
Cognitive Systems Monographs 38, https://doi.org/10.1007/978-3-319-97841-3_7

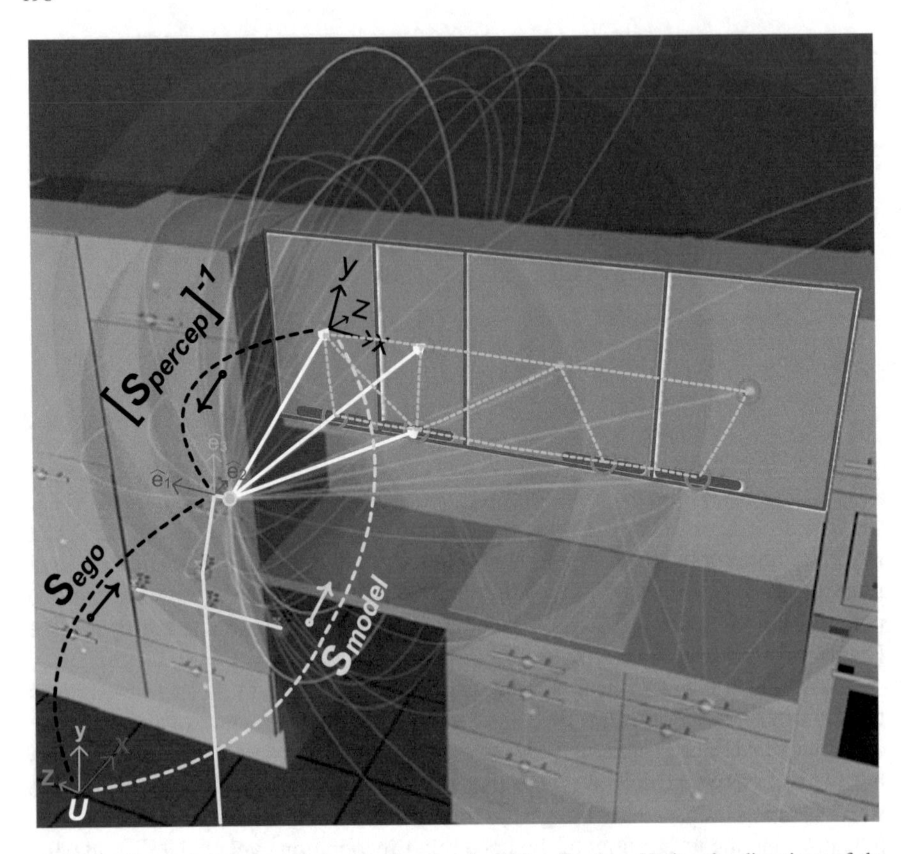

Fig. 7.1 Kinematic frames involved in the visual self-localization. Notice the directions of the coupling transformations to reveal the ego-frame S_{ego}

instrument to attain the generation and validation of the ego-center location candidates of the robot. This intersection of geometric primitives can also be done by a computational hierarchy of classes. This means the geometric algebra is not the only mathematical approach but it is a very convenient one due to its powerful set of operators and simplistic notation. Independently from the computational implementation, the treatment of subspaces helps to reduce the complexity of the visual-feature to shape-primitive matching by a computationally efficient, conceptually clear and consistent method for expressing the intersection among geometric-primitives.

7.1 Outline of Global Visual Self-Localization

This chapter presents a novel geometric and probabilistic approach for model-based visual global self-localization for humanoid robots. The global character of the localization concerns about the 6D-pose of the robot which can also be stated as the kidnapped robot problem [3]. The common approaches for global visual localization in task-specific environments (previously discussed in Chap. 2) are based on Bayesian filtering. Their initial state (prior on state distribution) is actually the tasks of global localization. This is why the approach of this work differs from the widespread prediction-updating recursive filtering schema. In this work, the global localization is split into two sequential phases; (i) The acquisition of visual-features also called landmarks. (ii) The simultaneous partial association between visual-features and world model representation for estimation of the 6D pose of the robot.

Visual Acquisition of Landmarks

The robust and accurate image acquisition (Chap. 4) for environmental visual object recognition (Chap. 5) are responsible for delivering the 6D-pose of elements (geometric visual-features) described in the world model (see Fig. 7.2). The visually recognized geometric-primitives (from now on geometric percepts) of the environmental objects provide not only useful information to perform actions but they also partially solve the data association between the visual and model spaces.

Simultaneous Partial Data Association and Pose Estimation

The extracted model geometric features of the scene impose restraints which are geometric-keys to deduct the pose of the robot. Each association of the form $\left\langle O_i^p, O_j^m \right\rangle$ constraints the position of the robot to the subspace of all points which are $||X_i^p||$ units away from X_j^m. This subspace is actually the surface of a sphere $\Omega\left\langle O_i^p, O_j^m \right\rangle$.

$$\underbrace{\Omega\langle O_i^p, O_j^m\rangle}_{\text{restriction subspace}} := X_j^m + \frac{1}{2}\underbrace{\left(||X_j^m|| - ||X_i^p||\right)}_{\text{perception-model matching}} e_\infty + e_0 \in PK^3, \tag{7.1}$$

centered at X_j^m with radius $||X_i^p||$ (see Fig. 7.3a). Note that the sphere in (Eq. 7.1) can also be represented as an element of the conformal geometric space PK^3, which has the Clifford algebra signature $\mathbf{G}_{(4,1)}$ [1]. For a single percept this idea provides little benefit, but on second thought, when observing the same concept with two different percepts it turns out to be a very profitable formulation because the ego-center should reside in both constrained subspaces.

This means that the ego-frame has to be on the surface of both spheres at the same time. Now consider two spheres acting as restriction subspace simultaneously constraining the position of the robot, $\Omega_1\left\langle O_i^p, O_j^m \right\rangle$ and $\Omega_2\langle O_k^p, O_l^m\rangle$, they implicate that the position of the robot belongs to both subspaces. Thus, the restricted subspace is a circle, namely an intersection of the two spheres (see Fig. 7.3b),

$$Z_{(1\wedge 2)} = \Omega_1\langle O_i^p, O_j^m\rangle \wedge \Omega_2\langle O_k^p, O_l^m\rangle. \tag{7.2}$$

Following the same pattern, a third sphere Ω_3 enforces the restriction to a point pair

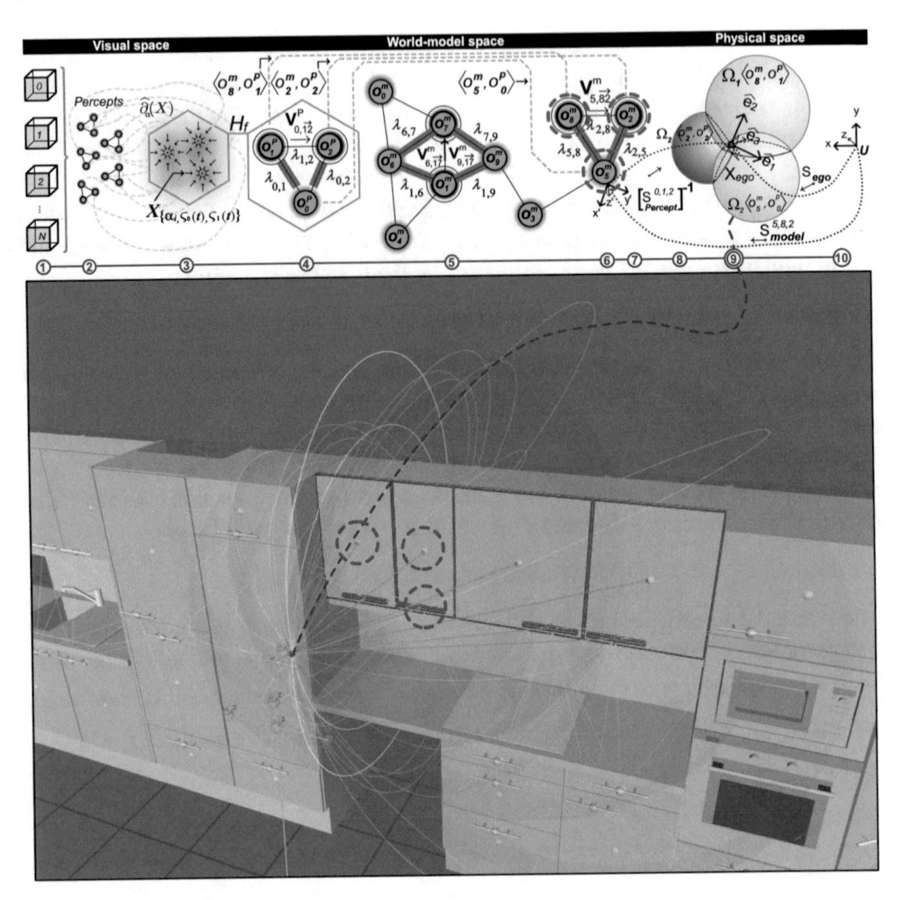

Fig. 7.2 Model-based visual self-localization approach (see [2]). From the schematic above: (1) Visual-feature extraction components (Chaps. 4 and 5). (2) Extracted geometric percepts mapped into the ego-frame. (3) Percepts. (4) Ego-percepts with corresponding model association hypotheses. (5) Percepts pruning upon world model. (6) Partial model matching. (7) Hypotheses generation. (8) Hypotheses validation. (9) Geometric and statistical pose estimation (Chaps. 6 and 7). (10) Resulting pose

$$J_{(1\wedge 2\wedge 3)} = Z_{(1\wedge 2)} \wedge \Omega_3 \left\langle O^p_r, O^m_s \right\rangle, \tag{7.3}$$

this is actually a circle-sphere intersection (see Fig. 7.3c). Finally, a fourth sphere Ω_4 determines the position of the robot, namely the intersection point from the this point pair (see Fig. 7.3d),

$$P_{(1\wedge 2\wedge 3\wedge 4)} = J_{(1\wedge 2\wedge 3)} \wedge \Omega_4 \left\langle O^p_t, O^m_h \right\rangle. \tag{7.4}$$

These concepts outline the position estimation idea in the proposed method. However the questions: (i) where to place? and (ii) how to set the radius of these spheres

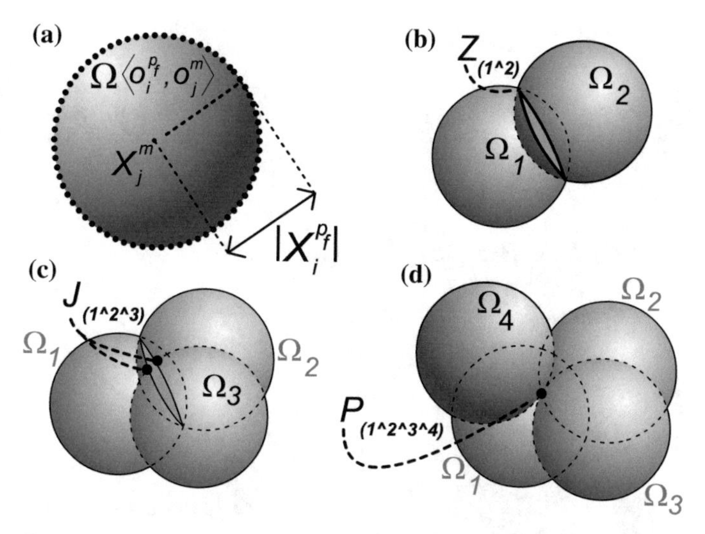

Fig. 7.3 **a** Geometric concept for robot position estimation. The surface of the sphere is the constraining subspace. **b** Co occurring constrained-subspaces depicting a circle. **c** Three constrained-subspaces acting in conjunction yield to a point pair. **d** Four constrained-subspaces yield to a simultaneity point, namely the point within the intersection of these four constrained-subspaces

in the world model? needs to be discussed: The partial association of the visual features acquired with the methodology of Chap. 5 allows to extract visual edges with euclidean metric, namely length and pose. These visual edges are ranked in terms of similarity according to their length using the method presented in Sect. 3.4.5. Once the ranked lists are created, the composition of queries helps to detect incidence between vertices and edges as explained in Sect. 3.4.5. This procedure generates partial matches between visual features and the world model. Now, partially matched geometric percepts of the world model are exploited in order to generate position candidates in terms of spheres as follows: First, the center point of the model edge O_j^m is use to place a restriction sphere Ω within the model space. The radius of this sphere is the distance measured during the depth estimation between the visual edge O_i^p and the camera center C. This is illustrated in Figs. 7.2-6, 8 and 7.4. Notice that the novel formulation and treatment of the uncertainty acquired during perception (based on the model presented in Chap. 6) is presented in Sect. 7.2. The computational complexity of this location hypotheses management process is $\frac{1}{2}n(n-1)(n-2)(n-3) = \frac{1}{2}n^4 - 3n^3 + \frac{11}{2}n^2 - 3n$ which is upper bounded by $O(n^4)$, where n is the cardinality of the subset of associations $\Omega\left\langle O_i^p, O_j^m\right\rangle$ (percepts-spheres). Fortunately, only in rare cases the internal partial result of the intersection stages are densely populated. This could be easily seen. For example, when intersecting two spheres, the resulting circle occupies a smaller subspace which in successive stages meets only fewer remaining spheres. One important factor why there are less operations in this combinational computation is because the child primitives resulting from the intersection of parent spheres should not be combined with their

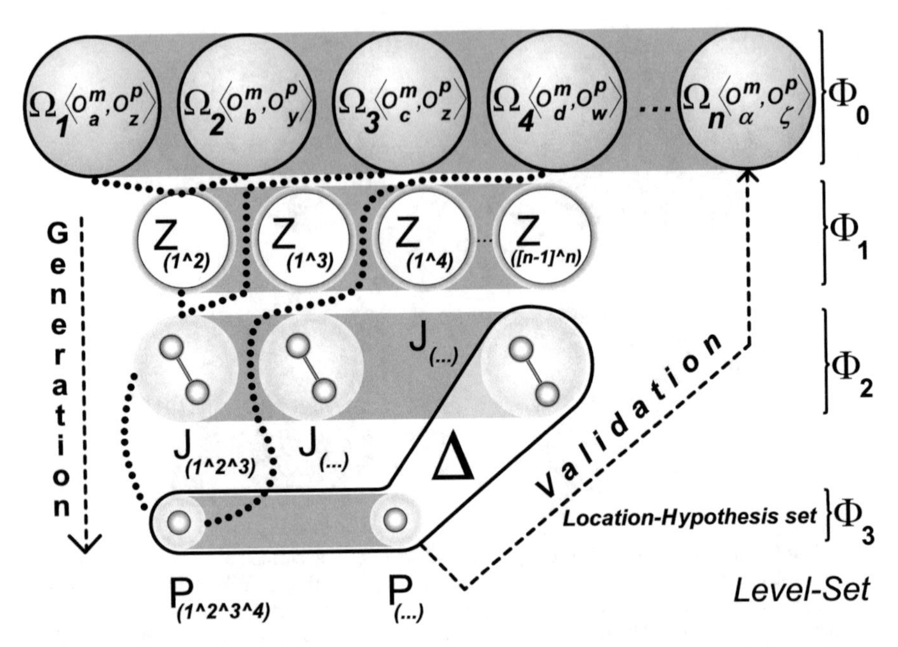

Fig. 7.4 Sphere intersection method for location hypotheses generation and validation systematically manages the location-hypotheses. In the upper part of the schema, the restriction spheres generated by partial association of model edges and visual edges constitute the first-level set. The intersection of this spheres generate the circles in the second-level set. In the next level, the intersection generates point pairs in the third-level set. Finally, the location-hypothesis set is produce by the intersection of point pairs and restriction spheres which are not related in previous levels. For example, this computational process requires 58140 intersection operations in the worst case when haven 20 spheres in the zero level set

relatives avoiding useless computation and memory usage, see this consideration in the schematic representation in Fig. 7.4.

Hypotheses Generation: Each percept subgraph is used to produce the *zero-level set*, composed of the first restriction spheres (see Fig. 7.2-7),

$$\Phi_0 = \left\{ \Omega_\zeta \left\langle O_i^m, O_j^p \right\rangle \right\}_{\zeta=1\ldots n}. \tag{7.5}$$

These spheres are intersected by either the *wedge* operator $\wedge$ (or well by the computational geometry class hierarchy) in an upper triangular fashion producing the *first-level set* Φ_1 containing circles. The *second-level set* Φ_2 is computed by intersecting those circles with spheres from Φ_0 excluding those directly above. The resulting point pairs are intersected in the same way creating the highest possible level (the *third-level set*) Φ_3, here the points resulting of the intersection of 4 spheres are contained. Finally, elements of Φ_2 which have no descendants in Φ_3 and all elements on Φ_3 represent location hypotheses

$$\Delta := \bigwedge_{\xi} \Omega_{\xi} \left\langle O_i^m, O_j^p \right\rangle. \tag{7.6}$$

Hypotheses Validation: Hypotheses are checked by selecting associations (see Fig. 7.2-8) $\left\langle O_i^p, O_j^m \right\rangle$ which were not considered in the generation of the current validating hypothesis. In case there is more than one prevailing hypothesis (which rarely happens in non-symmetric repetitive environments) a validation needs to take place selecting objects from the model and then localizing them in the visual space assuming the hypothesis 6D-pose. This is done in the same way that a particle filter determine its observation similarity, see [4].

Ideal Pose Estimation:

Once the location hypothesis has revealed, the position of the robot X_{ego} (see Fig. 7.2-10) and the orientation S_{ego} is expressed as

$$\underbrace{S_{ego}}_{self-localization} = \underbrace{S^{u,w,v}_{model}}_{model-matching} \underbrace{[S^{i,j,k}_{Percept}]^{-1}}_{visual-perception}, \tag{7.7}$$

which is actually the transformation from the kinematic chain that couples the world model frame S_{model} (forwards) and the perception frame $[S^{i,j,k}_{Percept}]^{-1}$ (backwards) (see Fig. 7.1). There are situations where a variety of diminishing effects alter the depth calculations of the percepts in a way that the ideal pose calculation may not be robust or could not be assessed. The subsequent sections describe the proposed method to determine the optimal location of the robot, namely the maximal probabilistic density position.

7.2 Geometry and Uncertainty Model

The critical role of the visual depth uncertainty can strongly diminish the precision of the estimated pose. It can also forbid the 6D-pose estimation by drawing away the intersection of the restriction subspaces. This means, the spheres might not intersect due to numerical instability and errors introduced by the visual perception. The geometric and probabilistic method for determining the position of the robot based on the intersection of spheres is properly formulated by introducing the following density sphere and its closed-form for optimal intersection.

7.2.1 Density Spheres

The proposed restriction spheres Ω_i are endowed with a density function

$$\widehat{f}(\Omega_i, x) \mid \Omega_i \in PK^3,\ x \in \mathbb{R}^3 \mapsto (0, 1] \in \mathbb{R}, \tag{7.8}$$

where PK^3 denotes the set of spheres in 3D space by means of conformal geometric algebra. The density value decreases as a function of the distance from any arbitrary point x to the surface of the sphere Ω_i

$$S(x, X_i, r_i) = |(||x - X_i|| - r_i)|, \tag{7.9}$$

$$\widehat{f}(\Omega_i, x) = \mathbf{exp}\left[\frac{-S(x, X_i, r_i)^2}{2\sigma_i^2}\right]. \tag{7.10}$$

This function depicts the non-normalized (by the factor $\frac{1}{\sigma\sqrt{2\pi}}$) radial normal distribution

$$\check{N}\left(\mu := \left\{x \mid \ker\left(S(x, X_i, r_i)\right)\right\}, \sigma_i^2\right), \tag{7.11}$$

for x to be in the surface of Ω_i, namely the null space of $S(x, X_i, r_i)$. Notice that the standard deviation σ_i refers to the model

$$\left[\mu_\Psi(|x|) := \sum_{i=0}^{n} a_i^\mu |x|^i \ , \ \sigma_\Psi(|x|) := \sum_{i=0}^{n} a_i^\Psi |x|^i\right] \tag{7.12}$$

presented in Chap. 6 at Sect. 6.7.2. The density of a point x in relation with a sphere Ω_i represents the non-normalized probability (density) for the point x to belong to the surface of the sphere Ω_i. The maximal density is obtained on the surface of the sphere itself. In order to estimate the robot position based on the intersection of spheres, it is necessary to propose an effective mechanism for applying intersection of density spherical subspaces for determining the robot position. The nature of the applied intersection has to consider the endowed spatial density of the involved density spheres. Concretely, the density intersection of spheres is determined by finding the subspace where the maximal density is located (see Fig. 7.5). This can be interpreted as an isotropic dilatation or contraction of each sphere in order to meet at the maximal density of the total intersection function (see Figs. 7.6 and 7.7), namely

$$\widehat{f}_t(x) \longrightarrow (0, 1] \in \mathbb{R}, \ x \in \mathbb{R}^3 \tag{7.13}$$

$$\widehat{f}_t(x) = \prod_i^n \widehat{f}(\Omega_i, x). \tag{7.14}$$

Due to the geometric composition of n spheres, it is possible to coarsely foresee the amount of peaks and the regions $\mathbf{W}_s$ where the density peaks are located (see Fig. 7.6c). Therefore, it is feasible to use state-of-the-art gradient ascendant methods [5] to converge to the modes using multiple seeds. These should be strategically located based on the spheres centers and intersection zones (see Fig. 7.6). Finally, the seed with maximal density represents the position

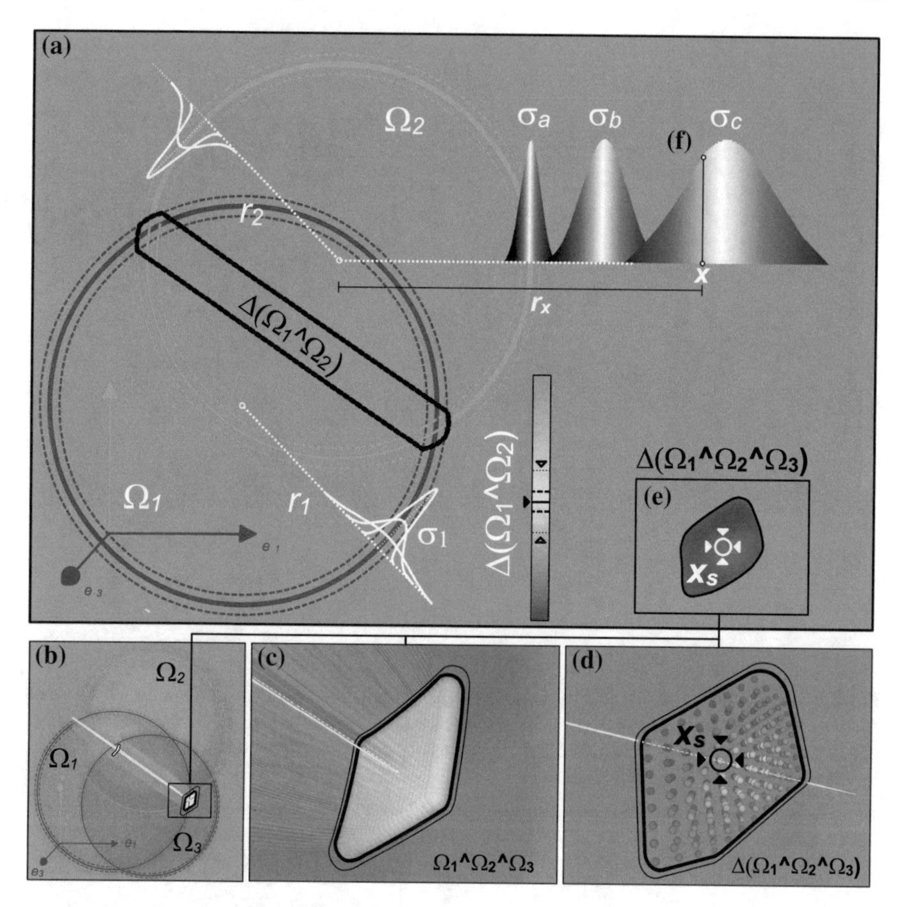

Fig. 7.5 Gaussian sphere meeting. **a** Two Gaussian spheres meeting $\Omega_1 \wedge \Omega_2$ describing a density-subspace $\Delta(\Omega_1 \wedge \Omega_2)$. **b** Three Gaussian spheres Ω_1, Ω_2 and Ω_3 meeting in two regions depicting a subspace $\Omega_1 \wedge \Omega_2 \wedge \Omega_3$. **c** Detailed view of one of the previous subspaces. **d** Discrete approximation of the maximal density location x_s. **e** Details of the implicit density-space $\Delta(\Omega_1 \wedge \Omega_2 \wedge \Omega_3)$. **f** (Upper-right) Implicit radius r_x when estimating the density at position x

$$x_s = argmax\ \widehat{f_t}(x). \tag{7.15}$$

However, there are many issues of this shortcoming solution commonly used in trilateration approaches [6]. The iterative solution x_s has a precision limited by the parameter used to stop the shifting seeds. In addition, the location and spreading of the seeds could have a tendency to produce undesired oscillation phenomena, under or oversampling and all other disadvantages of iterative methods. The optimization Eq. 7.15 can be properly solved in a convenient closed-form. In order to address the solution x_s, it is necessary to observe the configuration within a more propitious space which simultaneously allows a helpful representation of the geometrical constraint

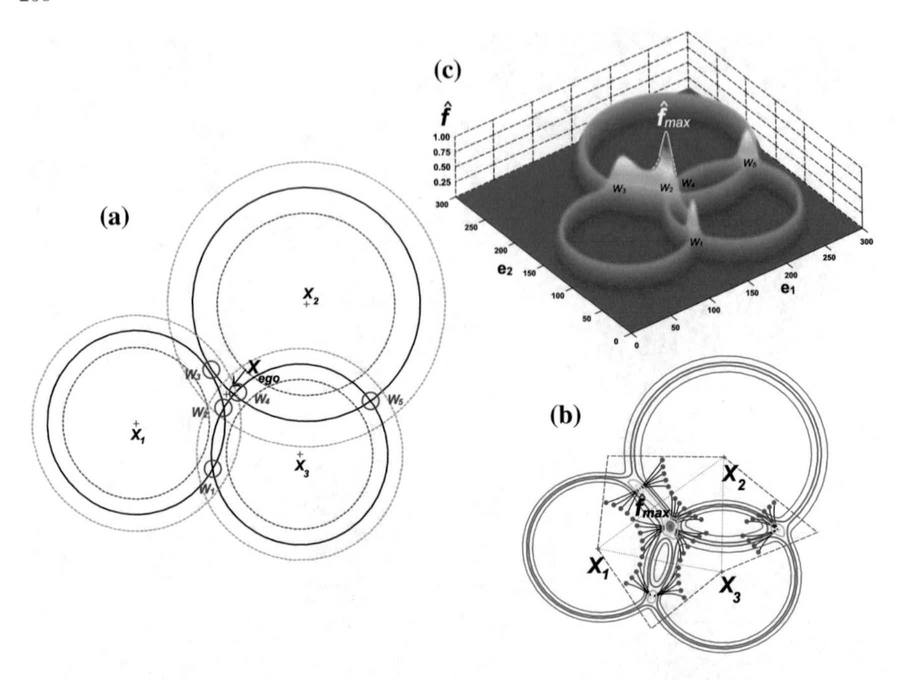

Fig. 7.6 Gaussian circles, namely 2D Gaussian spheres. **a** Three Gaussian circles setup. **b** The total accumulative density $\widehat{f}_c(x) = \sum_i^n \widehat{f}(\Omega_i, x)$ allows a better visualization of the composition of its product counterpart $\widehat{f}_t(x)$ (see also Fig. 7.7). **c** Density contours with seeds and their convergence by means of gradient ascendant methods

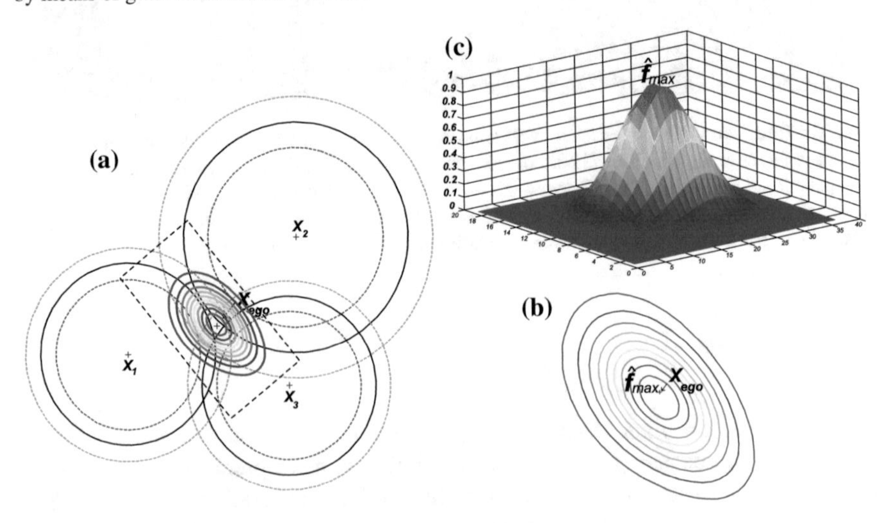

Fig. 7.7 The Gaussian circles, namely 2D Gaussian spheres. **a** Three Gaussian circles setup. **b** The total density $\widehat{f}_t(x) = \prod_i^n \widehat{f}(\Omega_i, x)$. **c** Density contours and ego-center X_{ego}. Notice that the resulting distribution is not Gaussian

and empowers an efficient treatment of the density. This has also to incorporate the measurements according to their uncertainty and relevancy to avoid density decay.

7.2.2 *Radial Space*

The key to attain a suitable representation of the optimization resides in the exponent of Eq. 7.10. The directed distance from a point x to the closest point on the surface of the sphere is expressed by Eq. 7.9. When considering the total density function in Eq. 7.14, the complexity unfolds expressing the total density as a tensor product. The inherent nature of the problem lies in the radial domain. The expression $S(x, X_i, r_i)^2$ is actually the square magnitude of the difference between the radius r_i and the implicit defined radius r_x between the center of the spheres X_i and the point x (Fig. 7.5f). Hence, the optimization can be better expressed in radial terms. The geometric constraints restricting the relative positions of the spheres in the radial space are presented in the following section.

7.2.3 *Restriction Lines for Sphere Intersection*

Consider the case of two spheres Ω_1 and Ω_2 (Fig. 7.8a). The radii of both spheres and the distance between their centers

$$\delta_{\overline{1,2}} = ||X_1 - X_2|| \tag{7.16}$$

allow the formulation of the geometric restrictions which ensure the intersection of the spheres in at least a single point P_χ. These restrictions are expressed by the inequation line L_χ describing the radial configuration subspace represented by radii pairs of the form $P_\chi = [r_1, r_2]^T \in \mathbf{S}^2$. This configuration represent the intersection of spheres $\Omega_1 \wedge \Omega_2$ where the $\mathbf{S}^2$ refers to the radial configuration space of spheres. Notice in Fig. 7.8d, the inequality line divides the configuration space into two regions. The half space holding the restriction imposed by the inequation line L_χ still contains configurations which produce no intersection of spheres. In fact any configuration holding

$$r_2 \geq \delta_{\overline{1,2}} + r_1. \tag{7.17}$$

In order to prevent these degenerated configurations two additional restriction inequation lines arise as follows.

Figure 7.8b shows the case where the minimal contact point P_β occurs, subject to

$$r_1 \geq \delta_{\overline{1,2}} + r_2. \tag{7.18}$$

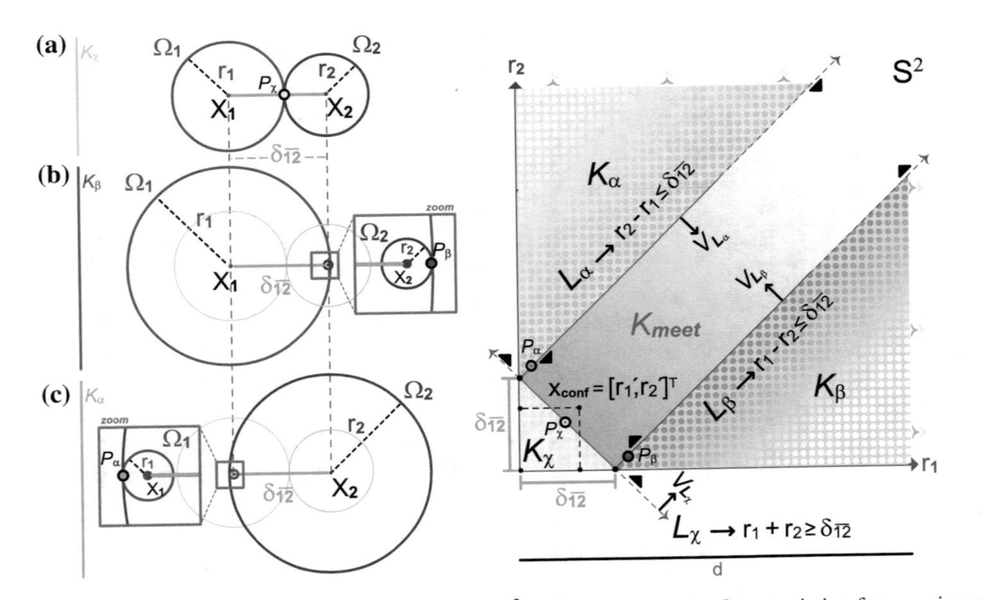

Fig. 7.8 The spheres intersection criteria by restriction lines in the radial space $\mathbf{S}^2$. **a** The line L_χ is the first restriction for ensuring non-empty intersection of spheres. **b** The derivation of right side non-empty intersection restriction line L_β. **c** The left side is the symmetric case generating the restriction Line L_α

In this configuration subspace, the sphere Ω_1 fully contains sphere Ω_2 and their surfaces intersect solely at P_β. In order to ensure at least one contact point, the variation of the radii of both spheres is restricted by a relation expressed by the inequality line L_β. The restriction actually applies in a symmetric manner by interchanging the roles from Ω_1 with Ω_2. This results in a third restriction, namely the inequality line L_α (see Fig. 7.8c, d). Therefore, the $\mathbf{S}^2$ space is divided in four regions K_α, K_β, K_χ and K_{meet} all open except K_χ. Only the configurations within the subspace K_{meet} represent non-empty intersections of the spheres, for example, the point x_{conf} in Fig. 7.8d with $x_{\text{conf}} = [r_1^{'}, r_2^{'}]^T \in K_{\text{meet}}$. The edge surface separating K_{meet} from other regions depict single point intersections of spheres, whereas elements within K_{meet} represent intersection depicting a circle with non-zero radius. This conceptualization soundly merges the distance among centers of the spheres with their radii. It produces a robust and general criteria to establish pairwise intersection guarantee (see Fig. 7.8d).

7.2.4 Restriction Hyperplanes for Sphere Intersection

The previous derivation of the restriction lines was achieved by considering only the case involving two spheres, however, it is possible to extend these restrictions to n spheres by considering few restrictions. Formally, this affirmation is theoretically supported by representing the n sphere radial configuration space $\mathbf{S}^n$ as the *Hilbert* space $\mathbf{C}^n$, where each dimension depicts the radius of one sphere. In element $x_{\text{conf}} \in \mathbf{S}^n$ of the n-dimensional radial configuration space can be uniquely specified by its coordinates with respect to orthonormal basis vectors $\hat{e}_i \in \mathbf{S}^n \mid i \in \{1, \ldots, n\} \subset \mathbf{Z}$, which are (as expected in a Hilbert space) perpendicular to each other because the radius of each sphere is independent from other radii. In this manner, the previous restriction lines could be perpendicularly extruded in $n-2$ dimensions creating the restriction hyperplanes $\Phi_\alpha^{(i,j)}$. Where each hyperplane divides the space in two subspaces. Configurations within the region opposite to the normal vector V_{L_α} (back of the hyperplane) represent non-intersecting spheres (see Fig. 7.9). Even more, the set of hyperplanes expressed in their Hesse normal form could be used to compose a matrix inequality $\mathbf{A}x \leq b$, where $\mathbf{A} \in \mathbb{R}^{m \times n}$ is a matrix, with m bounding half spaces separators (normal vectors of the hyperplanes) and b represents a $m \times 1$ column vector formed by stacking the Hesse distances of the hyperplanes, namely an open polytope (see Fig. 7.9). Consider the three spheres case (where $n = 3$) within the radial space each line $L_\alpha^{(i,j)}$, $L_\beta^{(i,j)}$ and $L_\chi^{(i,j)}$ could be extruded in the complementary dimension creating restriction planes given by $\Phi_\alpha^{(i,j)}$.

Next, the face cells, ridges and vertices of the polytope are found using a simple and fast implementation for vertex enumeration [7] (see Fig. 7.9). At this stage, it could be conveniently established whether the current configuration is valid, in other words, determine whether the point x_{conf} is inside the polytope. This assertion is formally given by

$$\mathbf{A}x_{\text{conf}} < b. \tag{7.19}$$

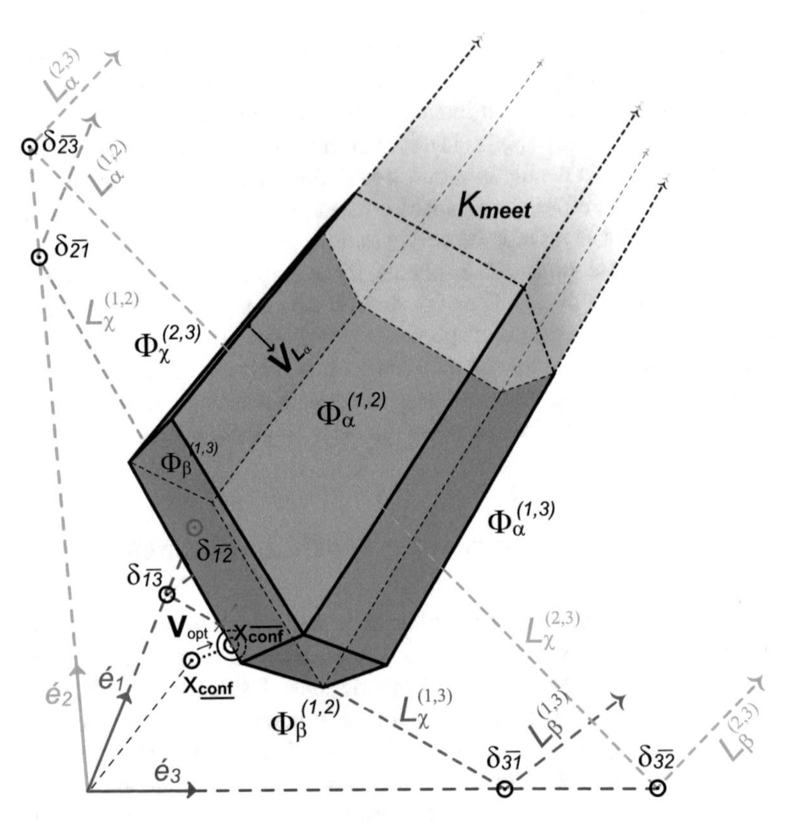

Fig. 7.9 The radial density space $\mathbf{Sb}^3$ containing the open polytope which delineates the subspace K_{meet}. The transformation-optimization vector V_{opt} implies an isotropic variation in the underlying density domain while creating a general dilatation within the implicit radial domain

In case this assertion is held (considering the perceptual depth correction of Eq. 6.6), there is no need to go through the following optimization phase because the spheres meeting on their surface, resulting the maximal density $\widehat{f}(x_{\overline{\text{conf}}}) = 1$. The opposite situations represent degenerated configurations resulting from noisy measurements and previously discussed uncertainties. For instance, the point $x_{\underline{\text{conf}}}$ represents an invalid configuration, outside of the polytope where no intersection of spheres exist (see Fig. 7.9). The solution for this case necessarily implies a decay in the density because at least one dimension (vector component) has to be modified for the point $x_{\underline{\text{conf}}}$ in order to become a valid configuration $x_{\overline{\text{conf}}}$. This offset signifies a dilatation or contraction of the sphere(s) depending on the magnitude and direction of the optimal displacement V_{opt}

$$x_{\overline{\text{conf}}} = x_{\underline{\text{conf}}} + V_{opt}, \tag{7.20}$$

which transforms the degenerated configuration into a valid one, namely a configuration of radii and distances among the spheres which produces a geometric and numeric stable intersection. The optimal criterion is to accomplish the minimal length offset vector transformation $V_{opt} := [v_{r_1}, \ldots, v_{r_n}] \in \mathbf{S}^n$. This minimal length is aimed in order to retain as much density as possible by eluding degradation of the spheres, namely reducing the radial deviation Eq. 7.10. The intuitive geometric way of finding such a vector is to find the closest point from $x_{\underline{\text{conf}}}$ on the cells or ridges of the polytope. This can be efficiently computed by projecting the point $x_{\underline{\text{conf}}}$ to each hyperplane

$$x_{\underline{\text{conf}}}^{(i,j)} = x_{\underline{\text{conf}}} - (V_\alpha^{(i,j)} \cdot x_{\underline{\text{conf}}})V_\alpha^{(i,j)}, \tag{7.21}$$

consequently selecting the closest point holding the assertion given by $\mathbf{A}x \leq b$. Although this technique is computationally efficient and geometrically correct the outcoming solution is not optimal. This occurs because within this space only the absolute directed distance is considered. No contribution effects of the different deviations (particular uncertainty profiles of each density sphere) are assessed, producing non-minimal density decay. This limitation could be overcome by considering a *homothety* transformation $\mathbf{H}(\mathbf{S}^n)$, namely a normalization of the radial configuration space inspired by the concept supporting the Mahalanobis distance [8]. The spatial density function of a Gaussian sphere Ω_i Eq. 7.10 could be conveniently reformulated in the radial domain as

$$\widehat{f}(\Omega_i, x) = \mathbf{exp}\left[-\frac{1}{2}\left(\frac{r_x}{\sigma_i} - \frac{r_i}{\sigma_i}\right)^2\right], \tag{7.22}$$

in such a way the deviation of the endowed normal distribution scales the implicit defined radius r_x and the mean radius r_i of the sphere Ω_i by the factor σ_i^{-1}. This normalization could be generalized for the whole radial configuration space $\mathbf{S}^n$ as

$$\mathbf{H} = diag\left[\sigma_1^{-1}, \ldots, \sigma_n^{-1}\right] \in \mathbb{R}^{nxn}. \tag{7.23}$$

This matrix actually represents the inverse covariance matrix Σ^{-1} of the total density function given by Eq. 7.24. This could be easily visualized by the alternative expression (by rewriting the exponent as a vector column and rearranging all in a standard form $x^T \Sigma^{-1} x$), namely

$$\widehat{f}_t(x) = \mathbf{exp}\left[-\frac{1}{2}\sum_{i=1}^{n}\left(\frac{||x - X_i||}{\sigma_i} - \frac{r_i}{\sigma_i}\right)^2\right]. \tag{7.24}$$

Based on Eq. 7.24 and taking into account the uncorrelated radial distributions, the underlying covariance matrix $\mathbf{H}^{-1} = \Sigma$ has zero elements outside its trace. Because of this fact, the proposed normalization $\mathbf{Sd}^n = \mathbf{H}(\mathbf{S}^n)$ could take place by applying the matrix $\mathbf{H}$ as an operator over the orthonormal vector bases of $\mathbf{S}^n$ as

$$\acute{e}_i = \mathbf{H}\hat{e}_i. \tag{7.25}$$

The euclidean metric within the resulting space $\mathbf{Sd}^n$ is uniformly isomorphic to the density space. Therefore, displacements of the same length arising from any position imply isotropic density decay in any direction reflecting different dilatation or contractions of those implicitly involved density spheres. Note, this normalization takes place before the vertex enumeration (for the polytope extraction) has been realized while computing the optimal points (Eq. 7.21, see Fig. 7.9). The application of the previous methods within the normalized radial configuration space $\mathbf{Sd}^n$ does not only ensure the optimal solution with minimal decay. It also benefits from the available certainty provided from those spheres with smaller deviation (closer and therefore higher reliable percepts) by introducing smaller displacements in their corresponding $\hat{e}_i$ dimension(s) in the displacement vector $V^d_{opt} \in \mathbf{Sd}^n$. In other words, the spheres which have a wider deviation can easily expand (or contract) their surfaces than those with smaller radius in order to obtain the highest possible intersection density at the meeting operation. This method delivers the optimal trade-off fusion while performing the management of the modeled uncertainty, see Fig. 7.10.

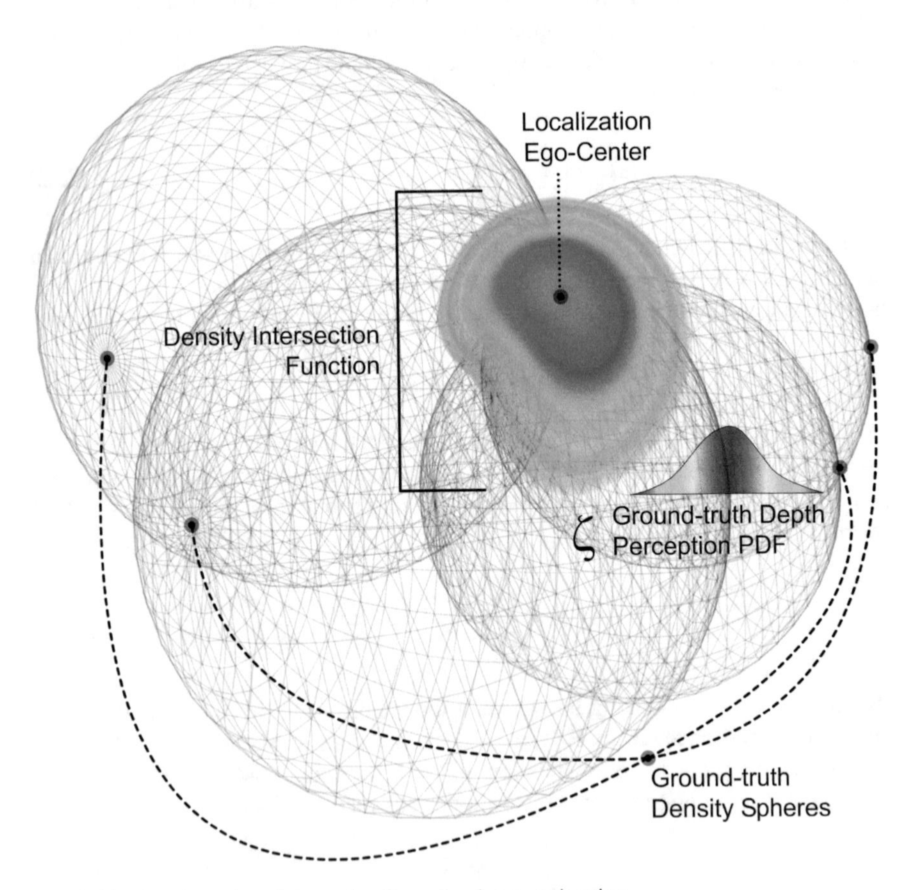

Fig. 7.10 Gaussian sphere intersection for optimal pose estimation

7.2.5 *Duality and Uniqueness*

In case the method has taken place in $\mathbf{Sd}^3$ (considering three spheres) for obtaining the optimal configuration $x_{\overline{\text{conf}}} \in \mathbf{Sd}^3$, still there is a duality to solve while back mapping the resulting radial configuration into the euclidean model space. This issue is directly solved computing the point pair

$$J_{\wedge_{i=1}^{3}} = \bigwedge_{i=1}^{3} \Omega_i \left(\sigma_i (x_{\overline{\text{conf}}} \cdot \hat{e}_i), X_i \right). \quad (7.26)$$

In case both solutions lie within the valid subspace a simple cross-check against the location of percepts which were not involved in previous calculations will disambiguate the solution. Furthermore, it is possible to obtain a unique solution by using four spheres for the optimization task, namely to represent the setup within $\mathbf{Sd}^4$. In this way $x_{\overline{\text{conf}}} \in \mathbf{Sd}^4$ could be again mapped back into the physical euclidean space by means of the meet operator unveiling the position of the robot as

$$P_{\wedge_{i=1}^{4}} = \bigwedge_{i=1}^{4} \Omega_i \left(\sigma_i (x_{\overline{\text{conf}}} \cdot \hat{e}_i), X_i \right). \quad (7.27)$$

7.3 Experimental Evaluation

This approach copes with the model-based visual self-localization using novel concepts from density spheres and computational geometry. The proposed method translates the probabilistic optimization problem of finding the maximal density location for the robot into a radial normalized density space $\mathbf{Sd}^n$ (Fig. 7.9). This allows a convenient description of the problem. Within this domain it is possible to determine the geometric restrictions which ensure the intersection of spheres. It also attains the optimal fusion and trade-off of the available information provided from the percepts by incorporating the available information of the landmarks according to their uncertainty. The global self-localization of the humanoid robot ARMAR-IIIa within the modeled environment was successfully performed using this approach. The scanning strategy takes 55–65 s processing 9 HDR stereo images, see Fig. 7.11. The model pruning takes 100–150 ms. The hypotheses generation-validation takes 270–540 ms. Finally, the vertex enumeration takes approximately 15–50 ms depending on the configuration, see Fig. 7.12. The experimental evaluation showed a maximal deviation of 64.8 mm at a distances of 2471.72 mm from the closest edge center in the environment.

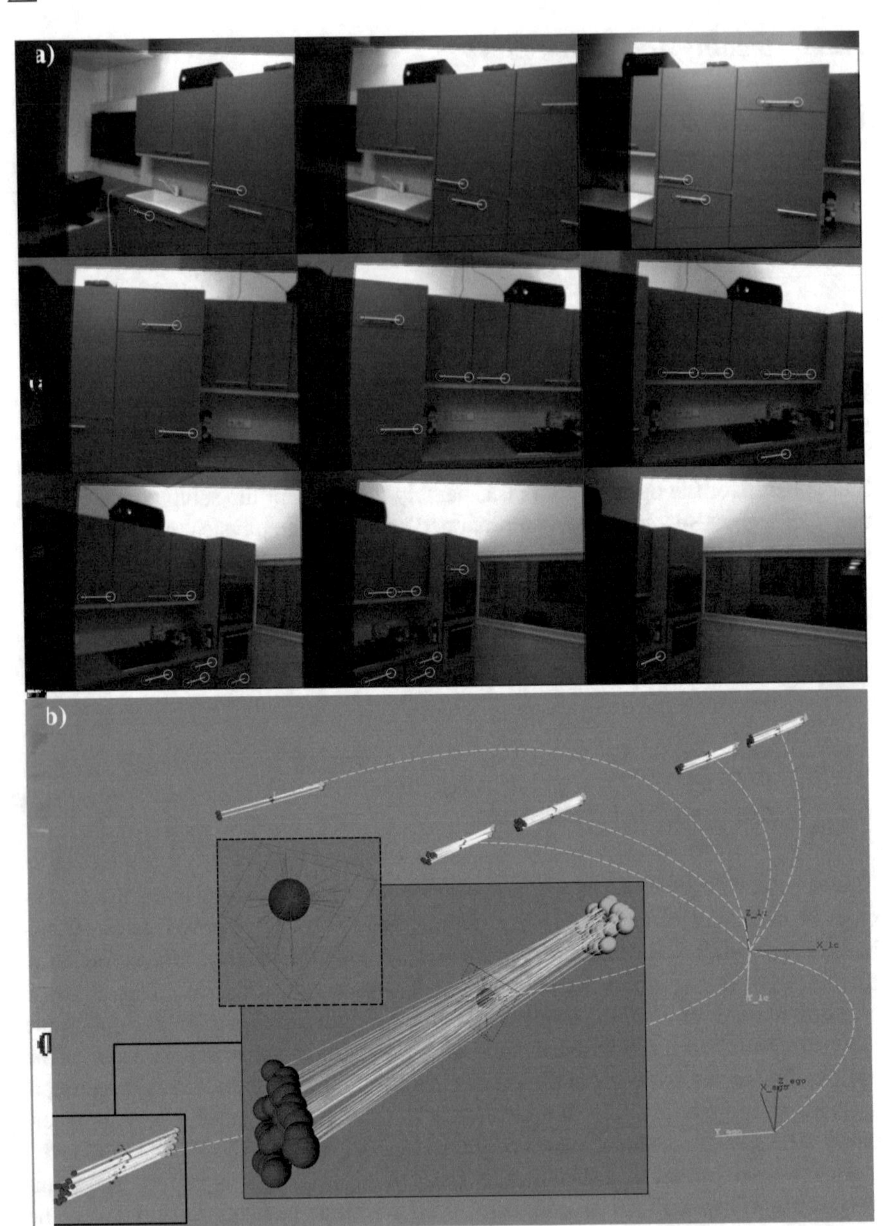

F ;. 7.11 **a** The acquisition of structural visual-features are landmarks in the context of visual global l alization. This images show the extraction of visual landmarks within a kitchen environment. T s elements are partially matched to the environment using their length attribute. **b** Visualization o he registered visual features. Notice the two frames: The cameras frame and the ego center frame l ated at the hip base of the humanoid robot ARMAR-IIIa

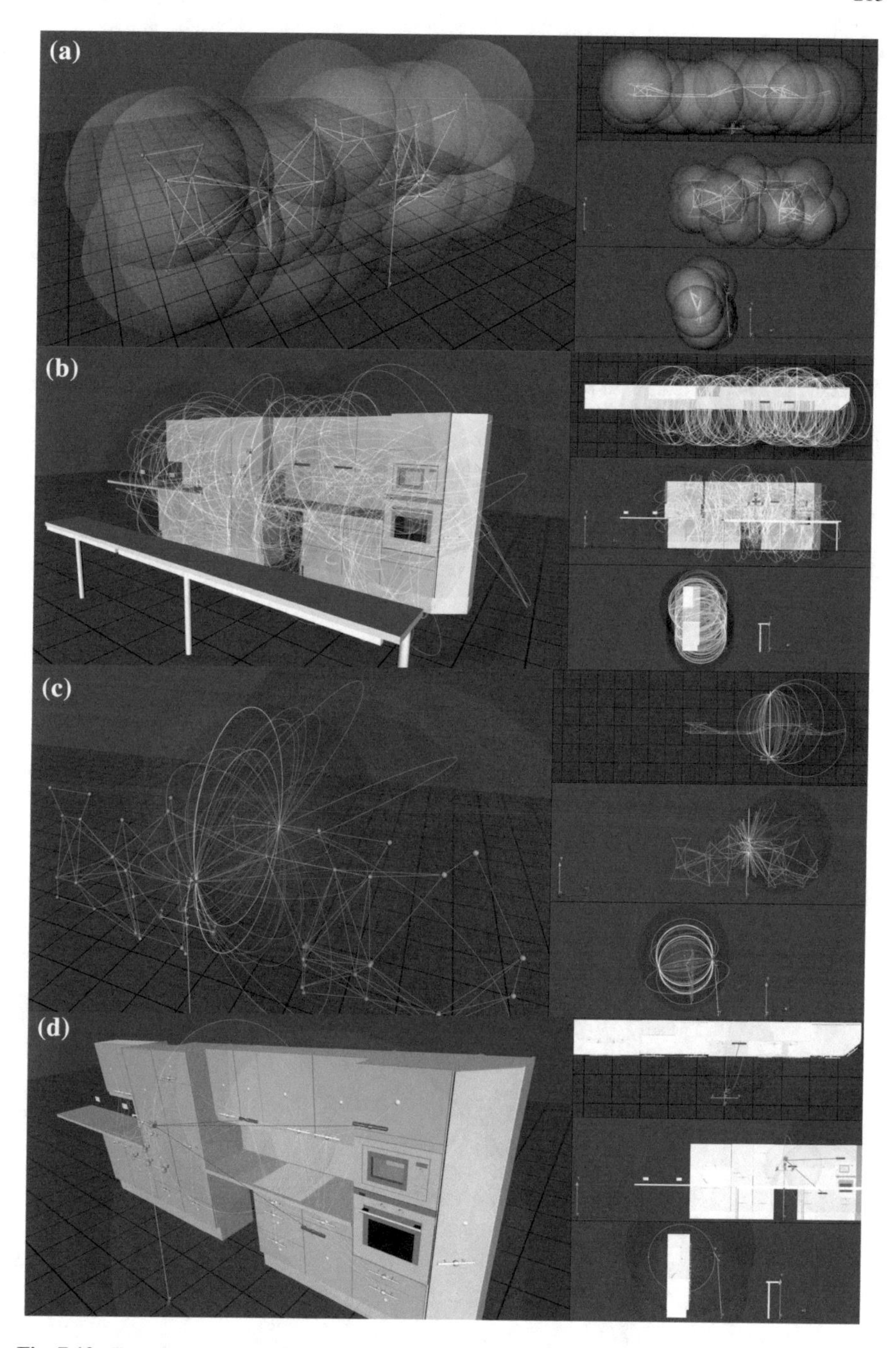

Fig. 7.12 Experimental trial of global visual localization. Results illustrate the simultaneous data association and pose estimation method and its implementation. **a** The generation of the first-level of restriction spheres. **b** The intersection of spheres generates the second-level composed by circles in 3D space. **c** After the intersection of the second and third level only few hypothesis remain. Spurious hypothesis are easily rejected by a cross-check

References

1. Bayro, E., and G. Sobczyk. 2001. *Geometric Algebra with Applications in Science and Engineering*. Birkhäuser. ISBN: 978-0-387-947464.
2. Gonzalez-Aguirre, D., T. Asfour, E. Bayro-Corrochano, and R. Dillmann. 2008. Model-based Visual Self-localization Using Geometry and Graphs. In *International Conference on Pattern Recognition*, 1–5.
3. Engelson, S., and D. McDermott. 1992. Error Correction in Mobile Robot Map Learning. In *IEEE International Conference on Robotics and Automation*, vol. 3, 2555–2560.
4. Dellaert, F., W. Burgard, D. Fox, and S. Thrun. 1999. Using the CONDENSATION Algorithm for Robust, Vision-based Mobile Robot Localization. In *IEEE Computer Society Conference on Computer Vision and Pattern Recognition*, vol. 2, 2588–2594.
5. Korn, T., and A. Granino. 2000. *Mathematical Handbook for Scientists and Engineeres*. Dover Publications. ISBN 978-0486411477.
6. Yu, Y.-B., and J.-Y. Gan. 2009. Self-localization using Alternating Combination Trilateration for Sensor Nodes. In *International Conference on Machine Learning and Cybernetics*, vol. 1, 85–90.
7. Avis, D., and K. Fukuda. 1991. A Pivoting Algorithm for Convex Hulls and Vertex Enumeration of Arrangements and Polyhedra. In *Annual Symposium on Computational Geometry*, 98–104.
8. Duda, R., P. Hart, and D. Stork. 2001. *Pattern Classification*, 2nd edn. New York: Wiley. ISBN 978-0471056690.

Chapter 8
Conclusion and Future Work

> *Great hopes make great men*
>
> Thomas Fuller

The main objective of this work is to establish the essential methods to couple the world model representation with the visual perception for humanoid robots in made-for-humans environments. The establishment of the vision-model coupling requires to achieve two central and correlated skills, namely the model-based visual environmental object recognition and the model-based visual global self-localization. The scientific contributions of this work are presented in Sect. 8.1. Finally, the future work is discussed in Sect. 8.2.

8.1 Scientific Contributions

The environmental model-based visual perception for humanoid robots consists of the processes which coordinately acquire, extract, organize, model and couple visual sensory information and world model representation in order to bidirectionally link signals and symbolic representations of the application domain. A solid and noticeable contribution to environmental model-based visual perception for intelligent autonomous humanoid robots operating in made-for-humans environments has been achieved in this research work. The scientific contributions were set at the critical aspects of visual sensing, feature extraction and pose estimation using the stereo vision system of humanoid robots within CAD-modeled environments.

Robust and Accurate Visual Manifold Sensing using Optimal Temporal and Exposure Fusion: The limitations imposed by the complex hardware composition of humanoid robots and the unsuitable environmental conditions (lighting and com-

D. I. González Aguirre, *Visual Perception for Humanoid Robots*,
Cognitive Systems Monographs 38, https://doi.org/10.1007/978-3-319-97841-3_8

plex material) in real application environments require novel methods for attaining robust and accurate visual manifolds from scenes. The proposed method overcomes these limitations providing high-dynamic-range images with high signal-to-noise ratio using standard low-dynamic-range cameras. The sensing method acts as accurate transducer between the physical world and the visual space by temporal image fusion and exposure bracketing. The temporal fusion developed in this work provides pixel-wise stability and robustness against arbitrary multimodal distributions of the noisy irradiance signals of the camera sensors. The exposure bracketing extends the dynamic range of the sensors by exploiting the extended and improved radiometric calibration coordinated by optimal exposure control in order to synthesize high-dynamic-range images with high quality for structural feature extraction. The statistical analysis of the stability and convergence of the fusion methods enables their synergistic coordination for noise removal and dynamic range enhancement. These statements are supported with experimental evaluation showing the high quality and applicability of the method.

Model-Based Visual Object Recognition using Saliency Eccentricity Graphs: The extraction of visual-features requires methods which significantly and coherently reduce the amount of information available in the visual manifold acquired by the sensing method. Previous methods in the literature for visual geometric-primitive extraction are conceptualized to select image structures based on thresholded saliencies and other parametric representations whose selection criteria are necessarily correlated with the image content. These content-dependent parameters can hardly be automatically estimated in a broadly manner. This limits the robustness and applicability of the geometric methods for object recognition. The proposed feature extraction and representation using subpixel saliency edge and rim graphs overcomes these limitations because the structural saliencies are selected based on their geometric constitution and not solely on their saliency magnitude. Hence, this novel method is free of image content parameters enabling wide applicability and high sensitivity. Another important feature of the method is the robust segmentation of geometric-primitives. This is done by the combined characterization of topological connectivity and spatial arrangement of the saliency subpixel graphs, namely the eccentricity model. As a result, the extraction and representation of geometric-primitives from HDR images expose no degeneration in presence of cluttered backgrounds and complex foregrounds. This is illustrated by the visual object recognition method which integrates 3D geometric-primitives from eccentricity graphs with the developed probabilistic world model representation. The experimental evaluation shows the high precision necessary for grasping and manipulation of complex environmental elements.

Model-Based Visual Global Self-Localization using Intersection of Density Spheres: Humanoid robots should attain their 6D-pose relative to the world coordinate system in order to plan sequences of actions necessary to realize useful tasks. This necessarily implies world model representations with task-specific information. In these representations, the environmental objects are registered relatively to the model coordinate system. Within these domains the visually extracted geometric features contain enough information to estimate the 6D-pose of the robot. This is realized by matching visual-features to shape-primitives. These two elements unambiguously define the kinematic chain linking the visual perception to the world model representation, namely the global localization in task-specific world model representations. The contribution of this work is the method which simultaneously solves both critical issues arising while creating this kinematic chain: (i) The association from visual-features to shape-primitives. (ii) The optimal pose assessment. The novelty of the contributions relies on two important aspects.

- **Visual depth uncertainty model**: The comprehensive analysis of the visual depth uncertainty was conducted based on reliable ground truth measurements. This precise data enables the exploitation of supervised learning methods in order to generate a novel visual uncertainty model without reductionism or assumptions of the visual perception process. The convenient model representation is computationally efficient not only for estimating the uncertainty distribution of the depth queries but also for correcting the systematic errors of the depth estimations.
- **Pose estimation by closed-form density spheres intersection**: A novel concept for data association and pose estimation based on multilateration of visual depth was proposed. This is realized by the intersection of probabilistic and geometric entities called density spheres. The particular depth-dependent density distribution of these spheres is the result of the uncertainty model. This enables the optimal integration of visual estimated geometric-primitives according to their uncertainty for the estimation of the robot position. The method is described by a closed-form solution for determining the position with maximal associated density resulting from the conjunctive combination of density spheres.

8.2 Future Work

The vision-model coupling is the first stage for environmental visual perception for humanoid robots. This bidirectional connection can be exploited for various important skills such as visual planning, navigation, dynamic localization, object registration for categorization, etc. In order to achieve these important visual skills, various scientific and technical questions should be addressed.

- **Active view fusion**: The comprehensive global visual localization and object recognition requires the active exploration of the surroundings in order to attain reliable recognition with 6D-poses. This implies multiview fusion to cope with situations where the initial field of view provides insufficient visual cues. Preliminary research in this direction has been done along this work [1]. However, there are still various important issues to be achieved for general active view fusion.
- **Environmental visual saliency learning**: An important performance improvement can be accomplished by enabling humanoid robots to learn elements or regions of the environment which are stable and salient for visual extraction and recognition. This research could be based on offline analysis using heuristic and photorealism render engines together with online unsupervised learning methods.

- **Real-time continuous global localization**: Based on the global localization method proposed in this work, it is possible to extend the state-of-the-art methods for dynamic localization in order to provide globally registered robot poses in real time. This implies novel methods to incorporate prediction of visual-features within the state estimation cycle. This is a fundamental skill for biped humanoid robots which solely rely on vision for obstacle avoidance, path planning and navigation.
- **Autonomous model generation**: The applicability of humanoid robots requires systems to autonomously generate environmental models. In this way, humanoid robots can act in various contexts without human assistance. This is a complex challenge beyond geometric or appearance representations including functional and semantic information.
- **Large scale applicability**: The environmental visual perception should cope with wide application scopes. Scenarios including full housing complexes and cities should be addressed for a full integration of humanoid robots in the society. These scenarios require to integrate diverse autonomous model generators producing shared and platform independent representations with different levels of granularity exploitable by multiple sensing approaches.
- **Beyond vision, multiple sensor fusion**: An unimpaired environmental perception for humanoid robots should also include other sensing modalities such as inertial measurement units (IMUs), global positioning system (GPS), etc. These sensors complement each other to cope with large scale applications.
- **Unified multiple purpose representation**: Intelligent autonomous humanoid robots require a unified representation of the environment, objects, actors (humans, animals, vehicles, other robots, etc.), internal aspects such as motion plans and events. Research in this directions is gradually bringing humanoid robots into everyday life.

Reference

1. Gonzalez-Aguirre, D., J. Hoch, S. Rohl, T. Asfour, E. Bayro-Corrochano, and R. Dillmann. 2011. Towards Shape-based Visual Object Categorization for Humanoid Robots. In *IEEE International Conference on Robotics and Automation*, 5226–5232.

Printed in the United States
By Bookmasters